Energy Conservation and Management for Professionals

RIVER PUBLISHERS SERIES IN ENERGY MANAGEMENT

Series Editors:

MICHELE ALBANO
Aalborg University, Denmark

The "River Publishers Series in Energy Management" is a series of comprehensive academic and professional books focussing on management theory and applications for energy related industries and facilities. Books published in the series serve to provide discussion and exchange information on management strategies, techniques, methodologies and applications, with a focus on the energy industry.

Topics include management systems, handbooks for facility management, safety, security, industrial strategies, maintenance and financing, impacting organizational communications, processes and work practices. Content is also featured for energy resilient and high-performance buildings.

The main aim of this series is to serve as a useful reference for academics, researchers, managers, engineers, and other professionals in related matters with energy management practices.

Books published in the series include research monographs, edited volumes, handbooks and textbooks. The books provide professionals, researchers, educators, and advanced students in the field with an invaluable insight into the latest research and developments.

Topics covered in the series include, but are not limited to:

- Facility management;
- Safety and security;
- Management systems and solutions;
- Industrial energy strategies;
- Financing and costs;
- Energy resilient buildings;
- Green buildings management.

For a list of other books in this series, visit www.riverpublishers.com

Energy Conservation and Management for Professionals

Benard Makaa
The Technical University of Kenya, Kenya

River Publishers

Routledge
Taylor & Francis Group
NEW YORK AND LONDON

Published 2025 by River Publishers
River Publishers
Alsbjergvej 10, 9260 Gistrup, Denmark
www.riverpublishers.com

Distributed exclusively by Routledge
605 Third Avenue, New York, NY 10017, USA
4 Park Square, Milton Park, Abingdon, Oxon OX14 4RN

Energy Conservation and Management for Professionals / by Benard Makaa.

Routledge is an imprint of the Taylor & Francis Group, an informa business

ISBN 978-87-7004-661-9 (hardback)
ISBN 978-87-7004-792-0 (paperback)
ISBN 978-87-7004-663-3 (online)
ISBN 978-87-7004-662-6 (master ebook)

Dedication

This book is dedicated to my wife Faith Chebet and my daughter Eve Wendo for their support and understanding during the countless hours spent on this book.

Contents

Preface

Welcome to this exploration of energy management and conservation, a field that sits at the crossroads of technology, policy, and sustainability. As the urgency to address climate change grows, energy efficiency and the use of renewable energy sources become even more important. This book is intended to serve as both a guide and an inspiration for professionals, students, policymakers, and anyone else interested in making a positive difference through positive energy practices.

Purpose of this Book

This book is intended to provide a comprehensive look at the strategies, technologies, and policies driving the transition to sustainable energy management. The content is designed to convey not only the technical aspects of energy systems, but also the socioeconomic and environmental imperatives that require a shift toward more sustainable practices. This book covers a wide range of topics relevant to current and future energy challenges, including the complexities of designing energy-efficient buildings and national energy policies.

Who Should Read this Book

- Energy professionals: Engineers, consultants, and managers working in the energy sector will gain advanced knowledge of system optimisation, project management, and sustainability integration.
- Policymakers and government officials: Those in positions of governance will gain a better understanding of the policies that can promote sustainable energy practices and how to effectively implement them.
- Academics and students: Educators and students of environmental science, engineering, and related fields will find useful information that complements and expands academic learning through practical and current applications.

- Environmentally conscious individuals: Anyone with a personal or professional interest in sustainability and modern energy challenges will find this book useful and informative.

What to Expect

The chapters of this book are designed to take the reader through both fundamental concepts and cutting-edge advancements in energy management. Each chapter not only delves into technical details, but also looks at case studies and real-world applications to provide a comprehensive picture of how energy conservation measures are being successfully implemented around the world.

Your Role

As you read these pages, remember that each of us has a role in shaping our energy future. Whether you are making decisions for a large corporation, establishing policies for a city, training the next generation of engineers, or simply looking to reduce your own environmental footprint, your actions contribute to a larger effort toward sustainability.

We hope that this book gives you the knowledge and confidence you need to take those actions, inspires you to come up with new solutions, and encourages you to lead by example. Let us look at how we can make a significant difference through conscious energy management and conservation.

Thank you for choosing to educate yourself on one of the most pressing issues of our time. Enjoy your journey through the pages of this book, and may it inspire you to take action and lead in your community toward a more sustainable future.

List of Figures

List of Tables

List of Abbreviations

AC	Alternating current
AMO	Advanced manufacturing office
APS	Announced pledges scenario
ASHRAE	American Society of Heating, Refrigerating and Air-conditioning Engineers
BEMS	Building energy management systems
BIM	Building information modelling)
BIPV	Building-integrated photovoltaic
BREEAM	Building research establishment environmental assessment methodology
Btu	British thermal unit
CEM	Certified energy manager
C&I	Commercial and industrial
CFL	Compact fluorescent lamp
COP	Coefficient of performance
Chu	Centigrade heat units
CHP	Combined heat and power
CNG	Compressed natural gas
DC	Direct current
DOE	Department of energy
DPP	Discounted payback period
ECI	Energy cost index
EECBG	Energy efficiency and conservation block grant
EGS	Enhanced geothermal systems
EJ	Exajoules
EMC	Energy management coordinator
EMS	Energy management systems
EnPI	Energy performance indicator
EnMS	Energy management systems
EPA	Environmental protection agency
EPS	Expanded polystyrene

EUI	Energy utilisation index
EV	Electric vehicle
GDP	Gross domestic product
GHG	Greenhouse gas
HEMS	Home energy management systems
HRV	Heat recovery ventilation
HVAC	Heating, ventilation, and air conditioning
IATF	International automotive task force
ICE	Internal combustion engines
IEA	International energy agency
IEE	Industrial energy efficiency
IEMS	Industrial energy management systems
IoT	Internet of things
IRENA	International Renewable Energy Agency
IRR	Internal rate of return
ISO	International Organization for Standardisation
LED	Light-emitting diode
LEED	Leadership in energy and environmental design
LNG	Liquefied natural gas
kW	Kilowatt
kWh	Kilowatt hour
kVA	Kilovolt-ampere
kVAR	Kilovolt-ampere reactive
NDC	Nationally determined contribution
NOI	Notice of intent
NPV	Net present value
NZE	Net Zero Emissions by 2050
PF	Power factor
PH	Potential of hydrogen
PPA	Power purchase agreement
RFI	Request for information
ROI	Return on investment
RPM	Revolutions per minute
SMART	Specific, measurable, achievable, relevant, time-bound
Solar PV	Solar photovoltaic
SPB	Simple payback period
STEPS	Stated policies scenario
TDM	Transportation demand management

TDS	Total dissolved solids
TES	Thermal energy storage
TOD	Transit-oriented development
TVM	Time value of money
VFD	Variable frequency drive
VIP	Vacuum-insulated panel
VRF	Variable refrigerant flow
VSD	Variable speed drive
WAP	Weatherization assistance programme
XPS	extruded polystyrene

1

Introduction

Chapter Preview

This chapter starts by asking and answering the questions, Why Energy Conservation and Management? Why Manage Energy? This section introduces the target audience. The book is written for professionals across diverse fields who are poised to make a significant impact in the realm of energy conservation and management. The book structure is also introduced.

1.1 Why Energy Conservation and Management Matter

In a world grappling with the twin challenges of resource depletion and climate change, energy conservation and management have never been more critical. This book addresses the global context of energy use, highlighting the pressing need to rethink our approach to energy resources. We delve into the implications of continuing on a path dependent on finite resources, underscoring how this strains our planet and threatens future energy security. The reality of climate change, driven in part by excessive energy consumption, especially from non-renewable sources, brings an urgent call to action.

For professionals in various fields, understanding and actively participating in energy conservation and management is not just a professional responsibility; it's a crucial part of being a global citizen. This book aims to equip professionals, from engineers to policymakers, with the knowledge and tools to make a tangible impact.

1. **Resource depletion and climate change**
    - Our planet faces two interconnected challenges: *resource depletion* and *climate change*.
    - **Resource depletion**: Our reliance on finite energy resources (such as fossil fuels) strains the earth's capacity to provide sustainably.

  - **Climate change:** Excessive energy consumption, particularly from non-renewable sources, contributes to global warming and its devastating effects.

2. **Finite resources and energy security**:
   - Continuing on a path dependent on finite resources jeopardises our future energy security.
   - **Energy conservation:** We reduce our dependence on scarce resources by using energy more efficiently.
   - **Management**: Effective energy management ensures optimal utilisation and minimises waste.

3. **Urgent call to action**:
   - Climate change is a reality driven partly by our energy choices.
   - We must act urgently to mitigate its impact and protect our planet.

4. **Global citizen responsibility**:
   - Energy conservation and management are not just professional responsibilities but essential for every global citizen.
   - Professionals, including engineers and policymakers, play a crucial role in shaping sustainable energy practices.

5. **Equipping professionals**:
   - This book aims to equip professionals with knowledge and tools to make a tangible impact.
   - By understanding energy dynamics and actively participating in conservation efforts, we contribute to a better future.

1.2 What Is Energy Conservation and Management

Energy conservation is increasingly becoming a topic of global importance. With the rapid depletion of natural resources and escalating environmental concerns, the significance of conserving energy cannot be overstated. Energy conservation involves reducing unnecessary energy usage to save resources and mitigate environmental impacts. It is not just about using less energy but using it wisely and efficiently. This concept is crucial for sustainable development, ensuring that future generations have access to necessary resources. We can collectively contribute to a healthier planet by understanding and implementing energy conservation strategies.

Energy conservation is the practice of reducing the use of energy. The primary goal is to maintain the same level of service without unnecessary or excessive energy use. It involves conscious efforts to minimise energy waste and optimise energy usage. This can range from simple actions like turning off lights when not in use to more significant initiatives such as upgrading to more energy-efficient appliances.

Energy conservation vs. energy efficiency

While often used interchangeably, energy conservation and energy efficiency are distinct concepts. Energy efficiency refers to using technology that requires less energy to perform the same function. For example, an energy-efficient LED bulb uses less power compared to a traditional incandescent bulb to produce the same amount of light. In contrast, energy conservation might involve using natural daylight instead of electric lighting when possible. Hence, while energy efficiency focuses on technology and equipment, conservation is more about behaviour and practices.

Why manage energy?

Managing energy is an essential practice for businesses, not only from an environmental standpoint but also for financial and operational reasons. Reducing energy consumption directly correlates with lowered operational costs and a decrease in greenhouse gas emissions, enhancing a company's environmental reputation. Furthermore, it provides a buffer against the unpredictability of energy prices and reinforces energy supply security by lessening reliance on imported energy.

However, implementing energy-saving measures often encounters hurdles. This could be due to a range of factors, including initial investment concerns, organisational inertia, or a lack of awareness about the potential benefits. Changes in organisational behaviour, which can significantly reduce energy usage, are sometimes overlooked due to these perceived barriers.

Adopting a systematic approach to energy management not only aligns with environmental stewardship but also offers significant benefits to an organisation:

Direct benefits

- **Energy cost savings:** Significant energy expense reduction directly impacts the bottom line.

- **Prioritisation of cost-effective energy saving opportunities:** Identifying and implementing no-cost and low-cost measures in daily operations to maximise efficiency.
- **Reduced greenhouse-gas emissions:** Lowering the ecological footprint through decreased emissions, contributing to environmental conservation.
- **Reduced exposure to changing energy prices:** Minimising the impact of fluctuating energy markets, leading to more predictable operational costs.
- **Reduced carbon footprint:** Demonstrating commitment to sustainability by decreasing the total amount of greenhouse gases produced.
- **Increased security of supply:** Lessening dependency on imported fuels enhances energy independence and stability.
- **Increased energy awareness among staff:** Cultivating a culture of energy consciousness, leading to increased employee engagement and participation in energy-saving practices.
- **Greater knowledge of energy use and consumption:** Gaining insights into energy patterns facilitates identifying areas for efficiency improvements.
- **Informed decision-making processes:** Utilising comprehensive data to make better strategic decisions regarding energy use.
- **Reduced uncertainty in future energy use:** Improved forecasting and understanding of future energy needs, allowing for more effective planning.

Indirect benefits

- **Positive publicity:** Gaining recognition for responsible energy practices, enhancing the public perception of the organisation.
- **Improved corporate image:** Strengthening brand reputation as an environmentally conscious and sustainable business.
- **Improved operational efficiencies:** Streamlining processes and systems for better productivity and reduced waste.
- **Improved maintenance practices:** Proactive maintenance leads to longer asset life and reduced downtime.
- **Improved safety and health:** Creating a safer workplace environment through better energy practices, potentially leading to reduced accidents and healthier work conditions.

1.3 The Target Audience for this Book

This book is primarily written for professionals across diverse fields who are poised to make a significant impact in the realm of energy conservation and management. Our target audience includes engineers, facility managers, energy consultants, urban planners, and policymakers. These professionals are at the forefront of implementing and advocating for energy-efficient practices and policies. The content is tailored to provide them with comprehensive insights, practical strategies, and real-world examples to effectively integrate energy conservation principles into their work and influence broader change.

It is crafted for professionals across various fields who are pivotal in driving energy conservation and management. It targets those who have the power and position to influence meaningful change in energy use and sustainability. From engineers to policymakers, this book aims to equip them with the knowledge and tools necessary for impactful action.

Engineers and their role in energy conservation

Engineers of all disciplines can find immense value in this book:

- **System design and innovation**: Engineers can utilise the principles outlined to design energy-efficient systems, from renewable energy technologies to green building designs.
- **Practical application**: Real-world examples in the book provide engineers with practical applications of energy conservation theories in their projects.

Facility managers and operational energy efficiency

Facility managers play a critical role in operational energy efficiency:

- **Building optimisation**: This book offers strategies for optimising energy use within buildings, focusing on systems like HVAC, lighting, and insulation.
- **Energy management**: Insights into energy management practices will help facility managers reduce energy costs and improve the sustainability of their operations.

Energy consultants as catalysts for change

Energy consultants can use this book to enhance their advisory roles:

- **Strategic planning and auditing**: The book provides frameworks for conducting energy audits and developing energy efficiency strategies for clients.
- **Guidance on sustainable solutions**: It serves as a guide for recommending the latest in sustainable energy solutions and practices.

Urban planners shaping sustainable cities

Urban planners will find the book beneficial in designing sustainable urban environments:

- **Sustainable urban design**: It covers integrating energy-efficient technologies and systems in urban planning.
- **Public transportation and infrastructure**: The book provides insights into planning energy-efficient public transportation and infrastructure.

Policymakers and the creation of energy policies

Policymakers can leverage this book to formulate effective energy policies:

- **Policy development**: It offers an understanding of how to develop policies that encourage energy efficiency and sustainability.
- **Regulatory frameworks**: The book highlights the importance of setting standards and incentives to promote sustainable practices.

Tailoring content for practical application and influence

The book is tailored to provide practical strategies and real-world examples:

- **Practical implementation**: It goes beyond theoretical knowledge, offering actionable strategies for professionals.
- **Influencing change**: The content empowers professionals to integrate energy conservation into their work and advocate for changes in their industries.

Uniting professionals in energy conservation efforts

This book serves as a unifying resource for professionals dedicated to energy conservation. It aims to foster a collaborative effort toward sustainable energy management by bridging diverse fields and creating a significant collective impact on global energy conservation.

1.4 Overview of the Book's Structure

The book is organised in fourteen well-articulated chapters on the nitty-gritty aspects of energy conservation and management that form the professional focus of the day. Thus, from the very basics of energy, environmental and economic impacts, energy audits, renewable sources, and policy regulations, the book chapters have been arranged in a way that each builds up on the previous ones to provide a holistic understanding of the issues. The book ends with some future trends, personal development for the professional, and an array of additional resources. This elaborate structure ensures that readers can not only get the hold of the essentials but also apply their knowledge effectively in real-world scenarios.

References

[1] IEA (2023), Energy Efficiency 2023, IEA, Paris https://www.iea.org/reports/energy-efficiency-2023, Licence: CC BY 4.0. (Accessed: 08 May 2024)

[2] IEA (2023), World Energy Outlook 2023, IEA, Paris https://www.iea.org/reports/world-energy-outlook-2023, Licence: CC BY 4.0 (report); CC BY NC SA 4.0 (Annex A). (Accessed: 08 May 2024)

[3] Barney L. Cape hart, Wayne C. Turner (2016), Guide to Energy Management, Fairmont Press, ISBN: 1420084895.

[4] Clive Beggs (2002), Energy: Management, Supply and Conservation, Butterworth Heinemann, ISBN: 0750650966

2

Understanding Energy and Its Importance

Chapter Preview

This chapter lays the basic foundation for understanding energy. It starts by describing the different types of energy in all its forms. We also explore the critical role of electricity in modern systems. We also introduce key metrics and units essential for understanding and communicating about energy. In addition, we also explore the current global energy consumption scenarios. The environmental and economic impact of energy consumption is also explored. We shall also discuss the economic implications of energy use, looking at how inefficiency in energy consumption can lead to increased costs for businesses and societies and how energy conservation can contribute to economic sustainability.

2.1 The Global Energy Landscape

Energy underpins much of modern life, from residential and industrial power to transportation fuel. It comes in a variety of forms and sources, each with a distinct significance in our everyday lives and the global economy. Understanding the many types of energy is critical for comprehending how our world works and the issues we confront in managing energy resources.

The global energy landscape is a complex and dynamic battleground, with various consumption patterns and varying reliance on energy sources.

Stated policies scenario (STEPS) sees lower demand projections for each of the fossil fuels. Total demand for fossil fuels declines from the mid-2020s by an average of 3 exajoules (EJ) per year to 2050 in the STEPS, and the peak in energy-related CO_2 emissions in the STEPS is brought forward to the mid-2020s (Figure 2.1).

The emergence of the new clean energy economy is starting to change the face of the energy system. For every USD 1 spent on fossil fuels, USD 1.8 is now being spent on clean energy; five years ago this ratio was 1:1 (Figure 2.2).

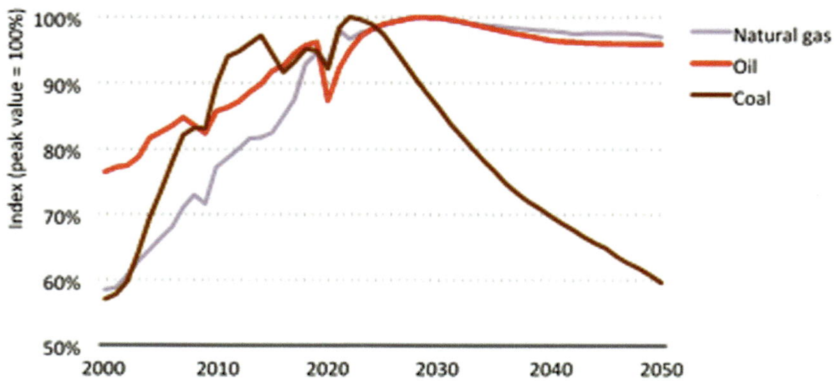

Figure 2.1 Fossil fuel consumption by fuel in the STEPS, 2000-2050 (Source, IEA, 2023 World Energy Outlook).

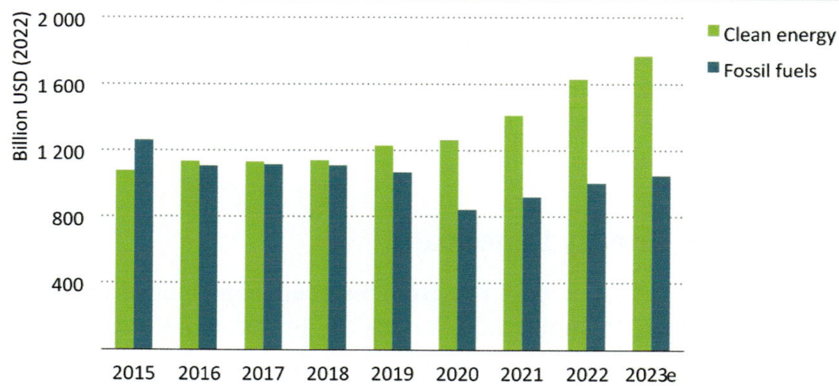

Figure 2.2 Global energy investment in clean energy and fossil fuels (Source, IEA, 2023 World Energy Outlook).

Notes: 2023e = estimated values for 2023. Numbers are in real 2022 US dollars.

Electricity demand rises over 80% to more than 150% by 2050 across scenarios and is met increasingly by low-emissions sources at the expense of unabated coal and natural gas (Figure 2.3).

Coal is the largest source of electricity in the world today, accounting for 36% of the total, but is overtaken by renewables by 2025 in all three scenarios (Figure 2.4). Solar PV capacity takes off in all scenarios, with only wind power at the same scale in the long term; their variable nature leads to increased deployment of battery storage

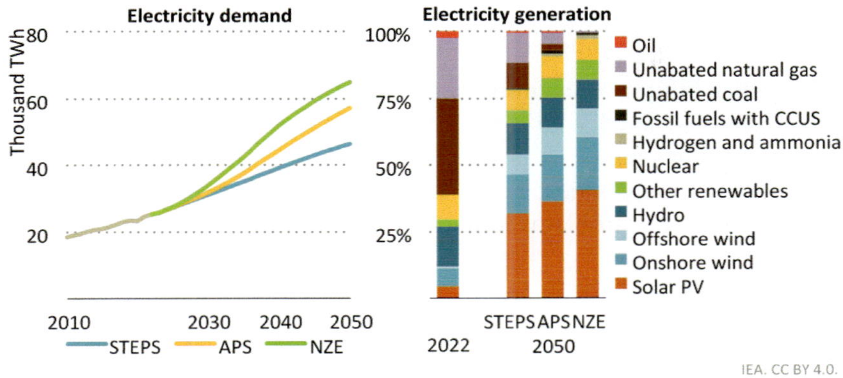

Figure 2.3 Global electricity demand, 2010-2050, and generation mix by scenario, 2022 and 2050.

Notes: TWh = terawatt-hours. Other renewables include bioenergy and renewable waste, geothermal, concentrating solar power, and marine power.

STEPS – Stated policies scenario, APS – Announced pledges scenario NZE – Net zero emissions by 2050.

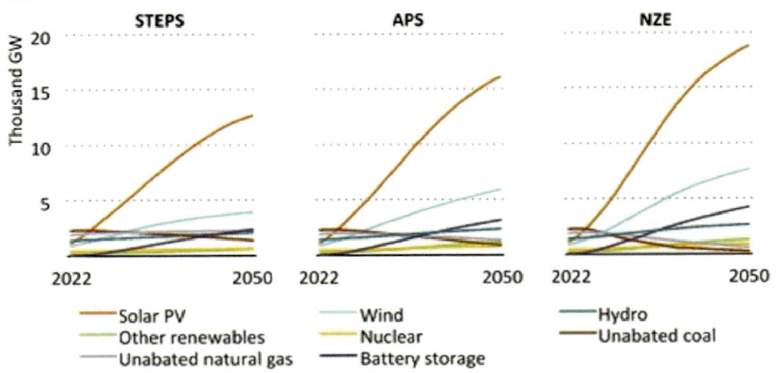

Figure 2.4 Global installed power capacity by selected technology and scenario, 2022-2050.

2.1.1 Current trends in global energy use

Recent research demonstrates diverse energy use habits around the world:

1. Energy source variability: While some regions rely mostly on oil and natural gas, others rely largely on coal or are fast embracing renewables.
2. Economic growth, industrialisation, and energy policy have a substantial impact on these consumption trends.

Reliance on non-renewable resources

The world's dependence on fossil fuels poses various challenges:

1. Supply instability and environmental impact: The use of fossil fuels is intimately associated with resource depletion, environmental damage, and climate change.
2. Geopolitical complexities: Fossil fuel production and trade frequently cause geopolitical problems, compromising global energy security and stability.

Renewables are playing an increasingly important role in the energy mix

1. The shift to renewable energy is gathering momentum.
2. Technological advancements: Increased solar and wind technology efficiency and cost reductions are making renewables more competitive.
3. Economic and policy drivers: Government incentives and lower costs are encouraging a move to renewable energy sources.

Challenges and opportunities for the energy transition

Transitioning to a renewable-based energy system entails various challenges:

1. Infrastructure and investment needs: Building renewable infrastructure and getting funding are huge difficulties.
2. Intermittency and grid integration: Managing the variability of renewable energy sources and incorporating them into existing systems are critical issues.
3. Opportunities for growth: This shift presents opportunities for economic growth, technical innovation, and environmental improvement.

2.2 The Environmental and Economic Impacts of Energy Consumption

Energy consumption is a major driver of environmental change and economic dynamics. Understanding its effects is critical for developing methods to mitigate negative consequences while maximising beneficial outcomes.

2.2.1 Environmental consequences of energy consumption

1. **Environmental consequences**:
   - Greenhouse gas emissions: Energy use, particularly from fossil fuels, contributes significantly to greenhouse gas (GHG) emissions. These emissions trap heat in the earth's atmosphere, contributing to global warming and climate change.
   - Pollution in the air is caused by the combustion of fossil fuels, which emit particulate matter, nitrogen oxides, and sulphur dioxide. These contaminants degrade air quality and endanger human health.
   - Deforestation and habitat loss: Energy generation frequently necessitates land use, resulting in deforestation and habitat degradation. Dam development for hydropower can disrupt ecosystems.
   - Water pollution: Some energy sources, such as coal mining, pollute water bodies with heavy metals and toxins.
2. **Link to climate change**:
   - Carbon dioxide (CO_2) is the principal GHG released by energy consumption. It adds to the greenhouse effect, which causes global warming.
   - Methane (CH_4) is emitted during fossil fuel extraction and transportation. It has a greater warming potential than carbon dioxide.
   - Nitrous oxide (N_2O) is emitted by agricultural operations and industrial procedures.

Greenhouse gas emissions and climate change

- Fossil fuels and emissions: The combustion of fossil fuels emits large amounts of greenhouse gases, which contribute considerably to global warming and climate change.
- Broader environmental impacts: In addition to emissions, energy use degrades air and water quality, destroys habitats, and contributes to biodiversity loss.
- Clean energy transition: Addressing these environmental challenges necessitates a move to cleaner, renewable energy sources.

Economic implications of energy use

1. **Economic implications**:
    - Costs of inefficiency: Inefficient energy usage increases costs for businesses and households. Squandered energy equals wasted money.
    - Energy prices fluctuate, which has an impact on economies. High energy costs can strain enterprises and reduce competitiveness.
    - Energy security: Relying on imported fossil fuels can expose economies to supply disruptions and price shocks.
    - Investment in renewable energy: Transitioning to renewable energy involves an initial investment but can result in long-term economic rewards.
2. **Energy conservation and economic sustainability**:
    - Efficient technologies: Using energy-efficient technologies lowers expenses while minimising environmental impact.
    - Renewable energy: Investing in renewable energy generates jobs, promotes innovation, and improves energy security.
    - Policy measures: Governments can encourage energy conservation through legislation, subsidies, and tax breaks.
    - Circular economy: Promoting circular behaviours (reuse, recycling, and trash reduction) saves energy and resources.

Inefficiency and costs

- Costs of inefficiency: Inefficient energy consumption increases operating costs and economic inefficiencies, affecting enterprises and communities.
- The economic burden of non-renewable energy: Dependence on non-renewable energy sources incurs additional costs, such as environmental clean-up and healthcare expenses.
- Efficiency benefits include cost savings, increased corporate competitiveness, and employment creation.

2.2.2 Energy conservation and economic sustainability

Impacts on sustainability

- Reducing costs and dependency: Energy conservation has an implied effect on reducing the costs of energy and limiting dependence on imported fuels.

- Stimulating economic growth: Investments in renewable energy and energy efficiency technologies indicate sustainable economic growth and job creation.

The balancing act: Environmental protection and economic growth

Challenges of balancing interests. It should be noted that the need for balance between environmental protection and economic development is challenging.

The role of integrated policies. Sustainable energy policies that integrate the stated environmental and economic goals make the best contribution to this balance.

Energy and the environment

The continued use of fossil fuels has environmental consequences, which is prompting a rethinking of how energy is consumed.

People in developing countries do not take energy for granted as much as people in developed countries do. Consider the number of everyday items, tools, and appliances that run on electricity – lamps, washing machines, televisions, radios, computers, and many other 'essential' items of equipment – all of which require a constant supply of electricity to function.

Imagine life without electricity. Our home and work lives would be very different. Indeed, our high-tech, computer-reliant society would cease to function; productivity would plummet, and GDP would be significantly reduced.

2.3 Politics and Self-interest

Any serious investigation into energy supply and conservation quickly reveals that the 'technical' aspects of the subject are inextricably linked to the 'politics' that surround it. This is because the two are inextricably linked; an available energy supply is the foundation of any economy, and politicians are keenly interested in how economies function. Politicians prefer short-term solutions and are hesitant to implement policies that will make them unpopular. Furthermore, many political parties receive funding from commercial organisations. Consequently, political self-interest frequently runs counter to collective reason.

The societal mind-set has a significant impact on energy conservation. Human nature has a significant impact on both politicians and consumers, but it does not always result in positive outcomes for society or the environment.

2.4 What Is Energy?

Let's examine a mass of 1 kg that has been lifted 1 m above its initial position. Evidently, the act of exerting force was necessary in order to raise the weight by a distance of one metre. Put simply, efforts have been made within the system to elevate the mass from a lower to a higher level. This work quantifies the quantity of energy that has been added to the system. When the weight is lifted, its energy level increases compared to when it is resting on the ground. Undoubtedly, this depiction is the foundation for the International System (SI) unit of energy, known as the 'joule', which can be precisely defined as:

One joule (J) is the work done when a force of 1 newton (N) acts on an object so that it moves 1 metre (m) in the direction of the force.

One newton (N) is the force required to increase or decrease the velocity of a 1 kg object by 1 m per second every second.

The number of newtons needed to accelerate an object can be calculated by eqn (2.1):

$$F = m \times a, \tag{2.1}$$

where m is the mass of the object (kg) and a is the acceleration (m/s^2). Given that the acceleration due to gravity is 9.81 m/s^2, a mass of 1 kg will exert a force of 9.81 N (i.e., 1 kg $\times$ 9.81 m/s^2). Therefore the energy required to raise it through 1 m will be 9.81 J.

When the 1 kg mass is released, it will descend a vertical distance of 1 metre and return to its initial position. By doing this, the 1 kilogramme mass will release the potential energy it has stored when it is at a higher level. It should be noted that the amount of energy released is equivalent to the amount of work exerted in lifting the weight. Therefore, the term 'work' is occasionally employed as a substitute for 'energy'.

Perhaps a good way of viewing energy is to consider it as stored work. Therefore, potential energy represents work that has already been done and stored for future use.

Potential energy can be calculated by eqn (2.2):

$$\text{Potential Energy} = m \times g \times h, \tag{2.2}$$

where m is the mass of the object (kg), g is the acceleration due to gravity (i.e., 9.81 m/s^2) and h is the height through which the object has been raised (m). As the weight falls it will possess energy because of its motion and this is termed *kinetic energy*. The kinetic energy of a body is proportional to its mass and to the square of its speed. Kinetic energy can be calculated by eqn (2.3):

$$\text{Kinetic energy} = 0.5 \times m \times v^2, \tag{2.3}$$

where v is the velocity of the object (m/s). We can see that during the time the mass takes to fall, its potential energy decreases whilst its kinetic energy increases. However, the sum of both forms of energy must remain constant during the fall. Physicists and engineers express this constancy in the 'law of conservation of energy', which states that the total amount of energy in the system must always be the same.

It should be noted that the amount of energy expended in raising the weight is completely independent of the time taken to raise the weight. Whether the weight is raised in 1 second or 1 day makes no difference to the energy put into the system. It does, however, have an effect on the 'power' of the person or machine performing the work. Clearly, the shorter the duration of the lift, the more powerful the lifter has to be.

Consequently, power is defined as the rate at which work is done, or alternatively, the rate of producing or using energy. The SI unit of power is the watt (W). Therefore, a machine requires a power of 1 W if it uses 1 J of energy in 1 second (i.e., 1 W is 1 J per second). In electrical terms, 1 W is the energy released in 1 second by a current of 1 ampere passing through a resistance of 1 ohm.

2.4.1 Units of energy

The most important types of energy are mechanical energy, electrical energy, and thermal energy. Different units have been assigned to various types of energy. However, it is important to note that because mechanical, electrical, and thermal energies are interchangeable, they can all be assigned the same unit.

***(i)* Mechanical energy.** The unit of mechanical energy is *newton-metre* or *joule* on the SI system.

The work done on a body is one newton-metre (or joule) if a force of one newton moves it through a distance of one metre, i.e. mechanical energy in joules = force in newton × distance in metres.

***(ii)* Electrical energy.** The unit of electrical energy is *watt-sec* or *joule* and is defined as follows:

One watt-second (or joule) energy is transferred between two points if a potential difference (p.d) of 1 volt exists between them and 1 ampere current passes between them for 1 second, i.e.
electrical energy in watt-sec (or joules) = voltage in volts × current in amperes × time in seconds.

Joule or watt-sec is a very small unit of electrical energy for practical purposes. In practice, for the measurement of electrical energy, bigger units *viz*., watt-hour and kilowatt hour are used.

$1 \text{ watt-hour} = 1 \text{ W} \times 1 \text{ hr} = 1 \text{ W} \times 3600 \text{ sec} = 3600 \text{ watt-sec}$

$1 \text{ kilowatt hour (kWh)} = 1 \text{ kW} \times 1 \text{ hr} = 1000 \text{ W} \times 3600 \text{ sec} = 36 \times 10^5 \text{ watt-sec}$.

(iii) **Heat.** Heat is a form of energy, which produces the sensation of warmth. The unit* of heat is calorie, British thermal unit (Btu) and centigrade heat units (Chu) on the various systems.

Calorie. It is the amount of heat required to raise the temperature of 1 g of water through 1°C, i.e. 1 calorie = 1 g of water × 1°C.

Sometimes a bigger unit namely *kilocalorie* is used. A kilocalorie is the amount of heat required to raise the temperature of 1 kg of water through 1°C, i.e.

$1 \text{ kilocalorie} = 1 \text{ kg} \times 1°\text{C} = 1000 \text{ g} \times 1°\text{C} = 1000 \text{ calories}$.

Btu. It is the amount of heat required to raise the temperature of 1 lb of water through 1°F, i.e. 1 Btu = 1 lb × 1°F.

Chu. It is the amount of heat required to raise the temperature of 1 lb of water through 1°C, i.e.

$1 \text{ Chu} = 1 \text{ lb} \times 1°\text{C}$.

*lb = Pound

Relationship among energy units

Energy, whether possessed by an electrical, mechanical, or thermal system, has one thing in common: it can do work. Mechanical, electrical, and thermal energies must all be measured in the same unit.

(i) Electrical and mechanical

1 kWh $= 1$ kW $\times$ 1 hr
$= 1000$ watts $\times$ 3600 seconds $= 36 \times 10^5$ watt-sec or joules
$\therefore$ 1 kWh $= 36 \times 10^5$ joules.
It is clear that electrical energy can be expressed in Joules instead of kWh.

(ii) Heat and mechanical

(a) 1 calorie $= 4.18$ joules (By experiment)

(b) 1 Chu $= 1$ lb $\times$ 1°C $= 453.6$ g $\times$ 1°C
$= 453.6$ calories $= 453.6 \times 4.18$ joules $= 1896$ joules
$\therefore$ 1 Chu $= 1896$ joules.

(c) 1 Btu $= 1$ lb $\times$ 1°F $= 453.6$ g $\times$ 5/9°C
$= 252$ calories $= 252 \times 4.18$ joules $= 1053$ joules
$\therefore$ 1 Btu $= 1053$ joules.

(iii) Electrical and heat

(a) 1 kWh $= 1000$ watts $\times$ 3600 seconds $= 36 \times 10^5$ joules

$$= \frac{36 \times 10^5}{4.18} \text{ calories} = 860 \times 10^3 \text{ calories}$$

1 kWh $= 860 \times 103$ calories or 860 kcal.

(b) 1 kWh $= 36 \times 105$ joules $= 36 \times 10^5/1896$ Chu $= 1898$ Chu
[1 Chu $= 1896$ joules]
$\therefore$ 1 kWh $= 1898$ Chu.

(c) 1 kWh $= 36 \times 10^5$ joules $= 36 \times 10^5$ $= \frac{36 \times 10^5}{1053}$
[1 Btu $= 1053$ joules]
$\therefore$ 1 kWh $= 3418$ Btu.

Kilowatt-hour (kWh)

The kilowatt-hour (kWh) is a particularly useful unit of energy which is commonly used in the electricity supply industry and, to a lesser extent, in the gas supply industry. It refers to the amount of energy consumed in 1 hr by the operation of an appliance having a power rating of 1 kW (eqn (2.4)).

Therefore:

$$1 \text{ kWh} = 3.6 \times 10^6 \text{ J}. \tag{2.4}$$

British thermal unit (Btu)

The British thermal unit (Btu) is the old imperial unit of energy. It is still very much in use and is particularly popular in the USA:

$$1 \text{ Btu} = 1.055 \times 10^3 \text{ J}. \tag{2.5}$$

Therm

The therm is a unit that originated in the gas supply industry. It is equivalent to 100,000 Btu:

$$1 \text{ therm} = 1.055 \times 10^8 \text{ J}. \tag{2.6}$$

Tonne of oil equivalent (toe)

The 'tonne of oil equivalent' (toe) is a unit of energy used in the oil industry:

$$1 \text{ toe} = 4.5 \times 10^{10} \text{ J}. \tag{2.7}$$

Barrel

The barrel is another unit of energy used in the oil industry. There are 7.5 barrels in 1 toe:

$$1 \text{ barrel} = 6 \times 10^9 \text{ J}. \tag{2.8}$$

Calorie

In the food industry the calorie is the most commonly used unit of energy. It is in fact the amount of heat energy required to raise 1 g of water through 1 °C:

$$1 \text{ calorie} = 4.2 \times 10^3 \text{ J}. \tag{2.9}$$

2.5 The Laws of Thermodynamics

Thermodynamics is the study of heat and work, and the conversion of energy from one form into another.

2.5.1 The first law of thermodynamics

The first law of thermodynamics is generally known as the law of conservation of energy. It states that the energy in a system cannot be created or destroyed. Instead, energy is changed from one form to another or moved between systems. The term 'system' can apply to anything from a basic object to a complicated machine. If the first law is applied to a heat engine, such as a

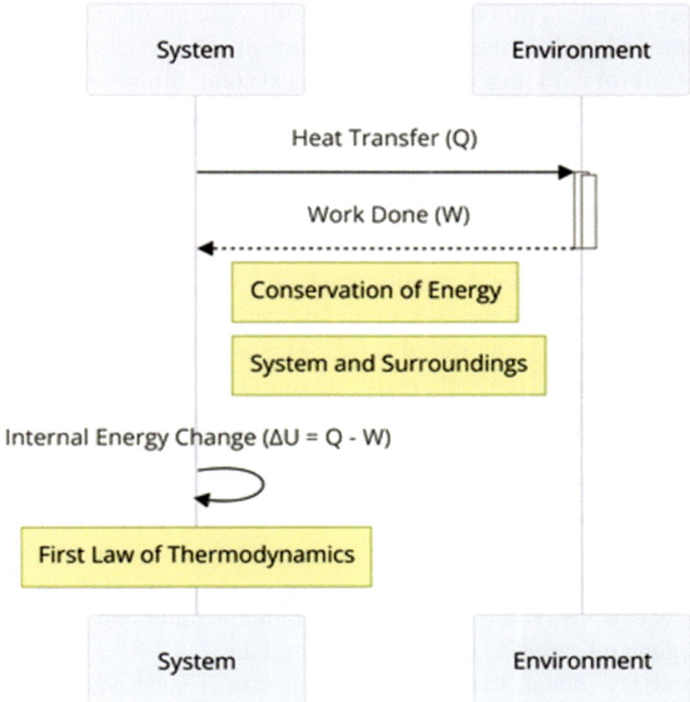

Figure 2.5 The first law of thermodynamics illustration.

gas turbine, where heat energy is turned into mechanical energy, it states that no matter how many stages are involved, the overall quantity of energy in the system must always be constant (Figure 2.5).

2.5.2 The second law of thermodynamics

The first law of thermodynamics describes the amount of energy in a system, but does not specify the direction of flow. The second law deals with the natural flow of energy processes. The second law of thermodynamics states that heat always flows from a hot object to a colder object (Figure 2.6(a)).

In another context, it explains why many natural processes behave as they do. For example, iron always rusts and never becomes pure iron. This is because all processes proceed in a way that increases the amount of disorder, or chaos, in the universe. Iron is produced by smelting ore in a foundry, which requires a significant amount of heat energy. So, when iron rusts, it goes back to a 'low-energy' state. Although it is a difficult concept to grasp, disorder has been quantified and dubbed 'entropy'. Entropy can be used to estimate the amount of useful work that can be done in a system. Simply put, the more chaotic a system, the more difficult it is to perform useful tasks.

In engineering, the second law of thermodynamics explains why a heat engine will never be completely efficient. Some of the heat energy from its fuel will be transferred to colder objects in the surrounding environment, preventing it from being converted into mechanical energy.

The second law of thermodynamics states that 'in all energy exchanges, if no energy enters or leaves the system, the potential energy of the state will always be less than that of the initial state'. This is also commonly referred to as entropy. A spring-driven watch will run until the potential energy in the spring is converted, and not again until energy is reapplied to the spring to rewind it. A car that has run out of gas will not run again until you refuel the car. Once the potential energy locked in carbohydrates is converted into kinetic energy (energy in use or motion), the organism will get no more until energy is input again. In the process of energy transfer, some energy will dissipate as heat. Entropy is a measure of disorder: cells are NOT disordered and so have low entropy. The flow of energy maintains order and life. Entropy wins when organisms cease to take in energy and die.

The second law states that there exists a useful state variable called entropy S. The change in entropy delta S (eqn (2.10)) is equal to the heat transfer delta Q divided by the temperature T.

$$\text{delta } S = \text{delta } Q/T \tag{2.10}$$

$$\Delta S = \text{Entropy} = \frac{\Delta Q}{T}.$$

For a given physical process, the combined entropy of the system and the environment remains a constant if the process can be reversed. If we denote the initial and final states of the system by 'i' and 'f':

$$S_f = S_i \ \text{(reversible process)}.$$

An example of a reversible process is ideally forcing a flow through a constricted pipe. Ideal means no boundary layer losses. As the flow moves through the constriction, the pressure, temperature, and velocity change, but these variables return to their original values downstream of the constriction. The state of the gas returns to its original conditions and the change of entropy of the system is zero. Engineers call such a process an isentropic

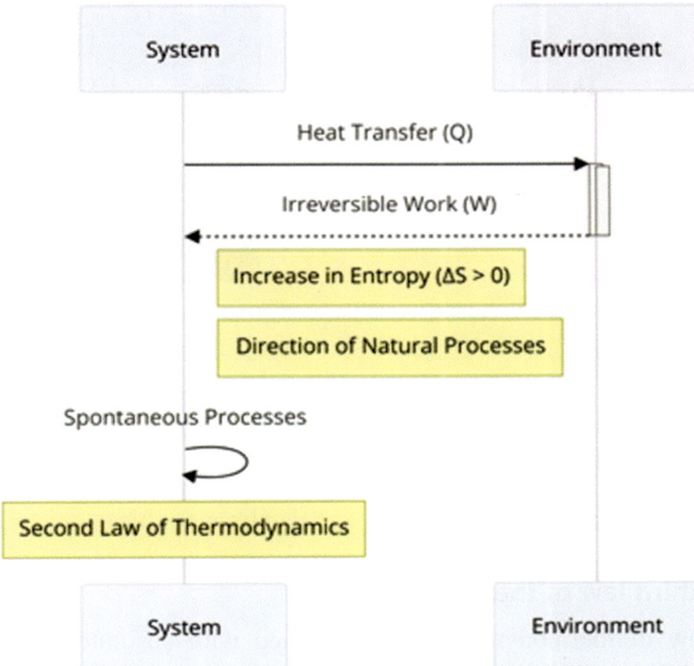

Figure 2.6(a) The second law of thermodynamics illustration.

process. Isentropic means constant entropy. The second law states that if the physical process is irreversible, the combined entropy of the system and the environment must increase. The final entropy must be greater than the initial entropy for an irreversible process:

$$S_f > S_i \text{ (irreversible process).}$$

An example of an irreversible process is a hot object in contact with a cold object. Eventually, they both achieve the same equilibrium temperature. If we then separate the objects they remain at the equilibrium temperature and do not naturally return to their original temperatures. The process of bringing them to the same temperature is irreversible.

$$\Delta S = \text{Entropy} = \frac{\Delta Q}{T}.$$

There exists a useful thermodynamic variable called entropy (S). A natural process that starts in one equilibrium state and ends in another will go

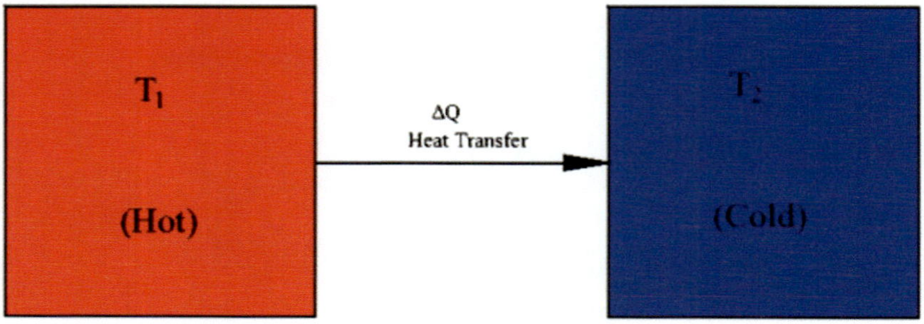

Figure 2.6(b) The second law of thermodynamics illustration.

in the direction that causes the entropy of the system plus the environment to increase for an irreversible process and to remain constant for a reversible process (Figure 2.6(b)).

$$S_f = S_i \text{ (Reversible)} \quad S_f > S_i \text{ (Irreversible)}$$

2.5.3 The third law of thermodynamics

The third law of thermodynamics is concerned with absolute zero (i.e., –273 °C). It simply states that it is impossible to reduce the temperature of any system to absolute zero (Figure 2.7).

Figure 2.8(a)(b) depicts an ideal heat engine that clearly demonstrates the first and second laws of thermodynamics. Heat engines, such as internal combustion engines and gas turbines, use thermal energy to generate mechanical work. They accomplish this by leveraging the temperature difference between a hot 'source' and a cold 'sink'. Heat flows from the hot 'source' to the cold 'sink' and is converted into mechanical energy by the engine's working parts.

If it is assumed that no energy is stored, then by applying the first law of thermodynamics it is possible to write down an energy balance for the system:

$$W = Q_H - Q_L \tag{2.11}$$

where W is the mechanical work produced by the engine (J), Q_H is the heat absorbed from the high-temperature 'source' (J), and Q_L is the heat rejected to the low temperature 'sink' (J).

Similarly, the efficiency, η, of the heat engine can be expressed thus:

$$\eta = \frac{\text{work output}}{\text{work input}} = \frac{W}{Q_H} = 1 - \frac{Q_L}{Q_H}. \tag{2.12}$$

Because the respective heat flows are proportional to the absolute temperature of the hot 'source' and the cold 'sink', it is possible to express the efficiency of an ideal heat engine as:

$$\eta = 1 - \frac{T_L}{T_H}. \tag{2.13}$$

According to the second law of thermodynamics, heat must flow from hot to cold. Eqn (2.13) shows that if there is no temperature difference between the hot 'source' and the cold 'sink', heat cannot flow, resulting in zero engine efficiency.

A large temperature difference between the hot 'source' and cold 'sink' leads to increased heat flow and higher cycle efficiency.

The second law of thermodynamics is so broad that it can be applied to explain how communities and ecosystems on earth behave when they consume energy.

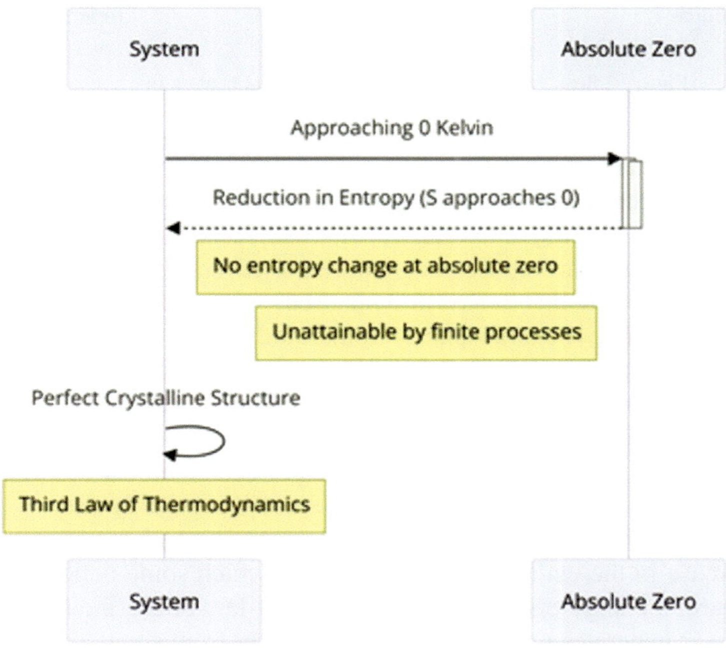

Figure 2.7 The third law of thermodynamics illustration.

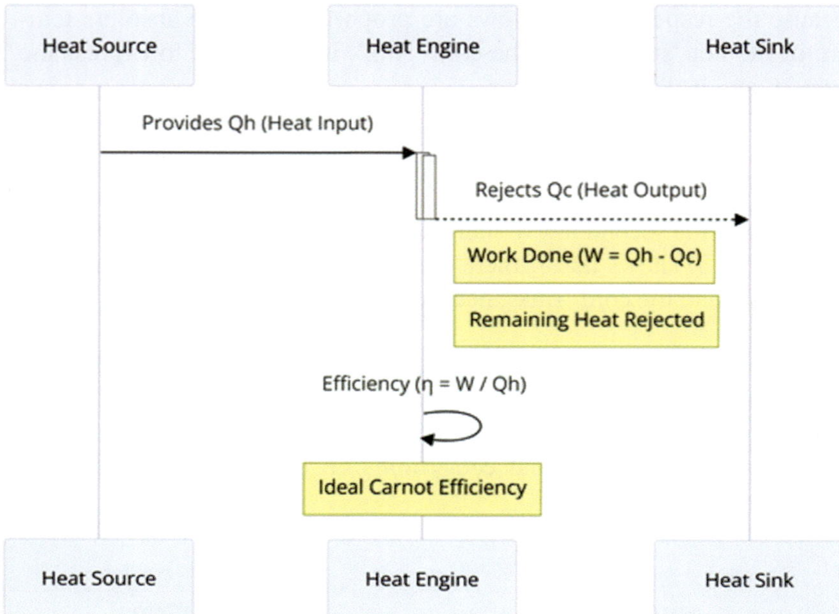

Figure 2.8(a) Schematic diagram of an ideal heat engine.

If environmental pollution is minimised and only renewable energy sources are used, the earth should remain relatively stable, allowing a low-entropy ecosystem to survive and thrive. If fossil fuels, such as petroleum, coal, and natural gas, are consumed, 'concentrated energy' from the sun, which has been stored in biomass for hundreds of thousands of years, is suddenly released into the atmosphere. In thermodynamic terms, the energy trapped in fossil fuels is highly ordered with low entropy. When burned, this highly ordered energy disperses into the environment, increasing its entropy, just as the second law of thermodynamics predicts. As more non-renewable fossil fuels are consumed, the second law predicts that entropy-related problems, such as pollution and global warming, will inevitably rise.

It is impossible to defy the second law of thermodynamics; entropy will always increase in the end! Even nuclear power, which some believe could solve the earth's energy crisis, follows the second law of thermodynamics. Nuclear power generates massive amounts of energy from small amounts of uranium. However, the second law states that once consumed, this highly ordered energy is dispersed into the environment, increasing overall entropy.

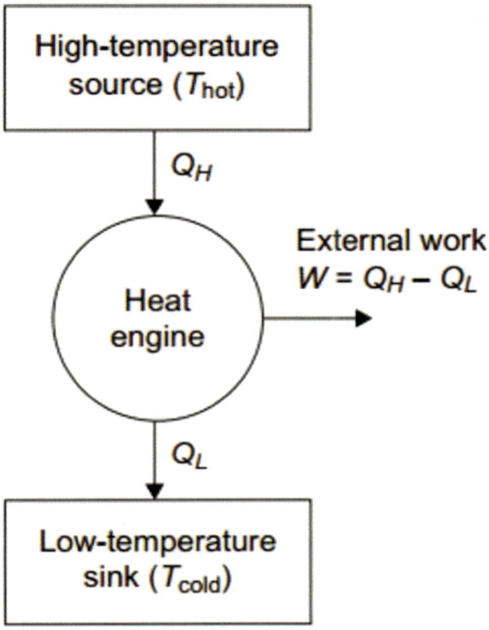

Figure 2.8(b) Schematic diagram of an ideal heat engine.

This increase in entropy may help to explain why safely disposing of nuclear waste has proven to be a significant challenge.

2.6 Electrical Energy

2.6.1 The role of electricity in energy systems

Electricity is central to modern energy systems:

Generation and distribution: Electricity is generated from a variety of sources and transmitted and distributed using power grids.

Storage challenges: Advances in battery technology and grid management are critical for dealing with the intermittent nature of renewable energy sources.

Electricity: An energy carrier

- Electricity is essential to our daily lives. It functions as an energy carrier rather than a primary energy source.

- Electricity is generated using various primary sources, including fossil fuels (coal, oil, and natural gas), nuclear energy, and renewable resources (solar, wind, and hydropower).
- Large-scale power plants generate electricity, which is then distributed via grids to homes, businesses, and industries.

2.6.2 Importance of electrical energy

Energy may be needed as heat, as light, as motive power etc. The present-day advancement in science and technology has made it possible to convert electrical energy into any desired form. This has given electrical energy a place of pride in the modern world.

Electrical energy is superior to all other forms of energy due to the following reasons:

(i) **Convenient form.** Electrical energy is a very convenient form of energy. It can be easily converted into other forms of energy. For example, if we want to convert electrical energy into heat, the only thing to be done is to pass electrical current through a wire of high resistance e.g. a heater. Similarly, electrical energy can be converted into light (e.g., electric bulb), mechanical energy (e.g., electric motors), etc.

(ii) **Easy control.** The electrically operated machines have simple and convenient starting, control, and operation. For instance, an electric motor can be started or stopped by turning on or off a switch.Similarly, with simple arrangements, the speed of electric motors can be easily varied over the desired range.

(iii) **Greater flexibility.** One important reason for preferring electrical energy is the flexibility that it offers. It can be easily transported from one place to another with the help of conductors.

(iv) **Cheapness.** Electrical energy is much cheaper than other forms of energy. Thus it is overall economical to use this form of energy for domestic, commercial, and industrial purposes.

(v) **Cleanliness.** Electrical energy is not associated with smoke, fumes, or poisonous gases. Therefore, its use ensures cleanliness and healthy conditions.

(vi) **High transmission efficiency.** The consumers of electrical energy are generally situated quite away from the centres of its production. The electrical energy can be transmitted conveniently and efficiently from

the centres of generation to the consumers with the help of overhead conductors known as transmission lines.

Energy can be used to generate heat, light, or motive power, among other things. Today's scientific and technological advancements have enabled the conversion of electrical energy into any desired form. This has made electrical energy a source of pride in the modern world.

Electrical energy outperforms all other types of energy for the following reasons:

1. Convenient form. Electrical energy is an extremely convenient source of energy. It is easily converted into other types of energy. For example, if we want to convert electrical energy into heat, we simply pass electrical current through a wire with a high resistance, such as a heater. Similarly, electrical energy can be converted into light (electric bulbs), mechanical energy (electric motors), and so on.
2. Simple control. Electric machines are simple to start, control, and operate. For example, an electric motor can be started or stopped by turning a switch on or off. Similarly, with simple arrangements, the speed of electric motors can be easily adjusted across the desired range.
3. Increased flexibility. Electricity is preferred due to its flexibility. It can be easily moved from one location to another with the help of conductors.
4. Cheapness. Electrical energy is much less expensive than other types of energy. This type of energy is cost-effective for residential, commercial, and industrial use.
5. Cleanliness. Electrical energy is not associated with smoke, fumes, or toxic gases. As a result, its use promotes cleanliness and health.
6. Excellent transmission efficiency. Consumers of electrical energy are typically located far from production sites. Electrical energy can be conveniently and efficiently transmitted from generation centres to consumers using overhead conductors known as transmission lines.

2.6.3 Generation of electrical energy

Electrical energy generation refers to the process of converting natural energy into electricity. Electrical energy is generated from naturally occurring energy sources. Electrical energy must be generated and delivered to the point of use at the precise moment it is required.

Energy can be obtained in a variety of forms from natural sources, including the pressure head of water, the chemical energy of fuels, the nuclear

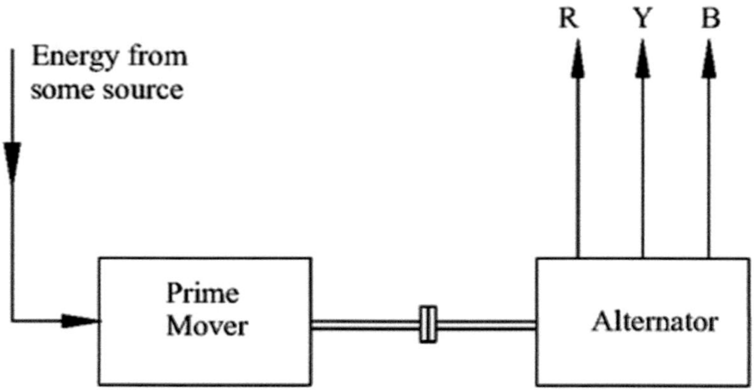

Figure 2.9 Schematic diagram of an ideal heat engine.

energy of radioactive substances, and so on. All of these types of energy can be converted into electrical energy by using appropriate arrangements.

The arrangement essentially employs (see Figure 2.9) an alternator coupled to a prime mover.

The prime mover is driven by the energy obtained from various sources such as burning of fuel, pressure of water, force of wind etc. For example, chemical energy of a fuel (e.g. coal) can be used to produce steam at high temperature and pressure. The steam is fed to a prime mover which may be a steam engine or a steam turbine. The turbine converts heat energy of steam into mechanical energy which is further converted into electrical energy by the alternator.

2.6.4 Sources of energy

Since electrical energy is produced from energy available in various forms in nature, it is desirable to look into the various sources of energy. These sources of energy are:

Because electrical energy is generated from energy found in various forms in nature, it is worthwhile to investigate the various sources of energy. The sources of energy are given below.

1. The Sun
2. The wind
3. Water
4. Fuels
5. Nuclear energy

Figure 2.10 PV system in a domestic situation.

Solar and wind energy are not widely used due to limitations. Currently, the other three sources, namely water, fuels, and nuclear energy, are primarily used to generate electrical energy.

I. The sun. The sun is our primary source of energy.
   There are two types of solar energy technology: photovoltaics (PV) and solar thermal. Solar PV, which you see on rooftops of homes and businesses, generates electricity directly from solar energy.
   Solar thermal technologies harness the sun's energy to generate heat, which is then converted into electricity. Reflectors allow the sun's heat energy to be focused over a small area. This heat can be used to generate steam, and electrical energy can be produced using a turbine–alternator combination. However, this method has limited applications because:

   - It requires a large area to generate even a small amount of electricity.
   - It cannot be used on cloudy days or at night because it is an inefficient method.

   Nonetheless, there are some areas of the world where strong solar radiation is received on a regular basis and mineral fuel sources are scarce or unavailable. Such locations are more appealing to solar plant builders.

II. The wind. This method can be used in situations where wind flows for an extended period of time. Wind energy is used to power a windmill, which drives a small generator. In order to continuously obtain electrical energy from a windmill, the generator is set up to charge the batteries. When the wind dies down, these batteries provide power. The advantages of this method include low maintenance and generation costs. However, this method has drawbacks such as:

- Variable output
- Uncertainty in wind pressure
- Low power generation

III. **Water.** When water is stored in a suitable location, it has potential energy due to the head created. This water energy can be converted into mechanical energy using water turbines. The water turbine powers the alternator, which converts mechanical energy to electrical energy. This method of generating electrical energy has gained popularity due to its low production and maintenance costs.

IV. **Fuels.** The primary sources of energy are fuels, specifically coal, oil, and natural gas. The heat energy of these fuels is converted into mechanical energy using appropriate prime movers such as steam engines, steam turbines, internal combustion engines, and so on. The prime mover powers the alternator, which converts mechanical energy to electrical energy. Although fuels remain the primary source of electricity generation, their reserves are depleting by the day. As a result, the current trend is to harness water power, which is a relatively constant source of power.

V. **Geothermal energy.** Geothermal energy is heat derived from the earth's subsurface. Water and/or steam transport geothermal energy to the earth's surface. Depending on its properties, geothermal energy can be used to heat and cool buildings or to generate clean electricity. However, electricity generation requires high or medium temperature resources, which are typically found near tectonically active regions. Geothermal energy accounts for a significant portion of Kenya's electricity demand.

VI. **Nuclear energy.** Toward the end of World War II, it was discovered that the fission of uranium and other fissionable materials releases a significant amount of heat energy. It is estimated that one kilogramme of nuclear fuel produces as much heat as 4500 tonnes of coal. With the proper arrangements, the heat produced by nuclear fission can be used to generate steam. The steam can power the steam turbine, which in turn drives the alternator to generate electricity. However, there are some challenges in the use of nuclear energy. The main challenges include:

Table 2.1 Comparison of sources of energy.

No.	Particular	Water power	Fuels	Nuclear energy
1	Initial cost	High	Low	Highest
2	Running cost	Less	High	Least
3	Reserves	Permanent	Exhaustible	Inexhaustible
4	Cleanliness	Cleanest	Dirtiest	Clean
5	Simplicity	Simplest	Complex	Most complex
6	Reliability	Most reliable	Less reliable	More reliable

- the high cost of nuclear plants;
- issues with radioactive waste disposal, and a lack of trained personnel to manage them.

The chief sources of energy used for the generation of electrical energy are water, fuels, and nuclear energy. Below is given their comparison in a tabular form (Table 2.1):

2.6.5 Efficiency

Energy efficiency, or system efficiency, is defined as the output energy divided by the input energy.

$$\text{Efficiency, } \eta = \frac{\text{Output energy}}{\text{Input energy}}.$$

As power is the rate of energy flow, therefore, efficiency may be expressed equally well as output power divided by input power i.e.

$$\text{Efficiency, } \eta = \frac{\text{Output power}}{\text{Input power}}.$$

Example 1. Mechanical energy is supplied to a d.c. generator at the rate of 4200 J/s. The generator delivers 32.2 A at 120 V.

i. *What is the percentage efficiency of the generator?*
ii. *How much energy is lost per minute of operation?*

Solution.

(i) *Input power, Pi* $= 4200$ *J/s* $= 4200$ *W*
Output power, Po $= EI = 120 \times 32.2 = 3864$ *W*
$\therefore$ *Efficiency,* $\eta = \frac{p_o}{p_i} \times 100 = \frac{3864}{4200} \times 100 = 92\%$

Table 2.2 Calorific value of fuels.

No.	Particular	Calorific value	Composition
1	**Solid fuels**	5000 kcal/kg	C = 67%, H = 5%, O = 20%,
	(*i*) Lignite	7600 kcal/kg	ash = 8%
	(*ii*) Bituminous coal	8500 kcal/kg	C = 83%, H = 5.5%, O = 5%,
	(*iii*) Anthracite coal		ash = 6.5%
			C = 90%, H = 3%, O = 2%,
			ash = 5%
2	**Liquid fuels**	11,000 kcal/kg	C = 86%, H = 12%, S = 2%
	(*i*) Heavy oil	11,000 kcal/kg	C = 86.3%, H = 12.8%, S = 0.9%
	(*ii*) Diesel oil	11,110 kcal/kg	C = 86%, H = 14%
	(*iii*) Petrol		
3	**Gaseous fuels**	520 kcal/m^3	CH4 = 84%, C2H6 = 10%
	(*i*) Natural gas	7600 kcal/m ^{3}	Other hydrocarbons = 5%
	(*ii*) Coal gas		CH4 = 35%, H = 45%, CO = 8%,
			N = 6%
			CO2 = 2%, Other hydrocarbons = 4%

(*ii*) *Power lost,* $PL = Pi - Po = 4200 - 3864 = 336\ W$

$\therefore$ *Energy lost per minute* (= 60 s) *of operation*

$= P_L \times t = 336 \times 60 = 20160\ J.$

2.6.6 Calorific value of fuels

The calorific value of a fuel is the amount of heat produced during combustion.

The calorific value of a fuel indicates how much heat it can generate. Fuel with a higher calorific value produces more heat. The calorific value of solid and liquid fuels is expressed in cal/g or kcal/kg, respectively. However, for gaseous fuels, it is generally stated in cal/litre or kcal/litre (Table 2.2).

2.6.7 Advantages of liquid fuels over the solid fuels

i. Liquid fuels are easier to handle and take up less storage space.
ii. The combustion of liquid fuels is uniform.
iii. Solid fuels have a higher moisture content, making them difficult to burn. Liquid fuels burn more easily and quickly than solid fuels.
iv. Solid fuels generate a large amount of ash, which can be difficult to dispose of.
v. However, liquid fuels produce no or little ash after combustion.

vi The firing of liquid fuels is easily controlled. This allows for easy adjustment to fluctuating load demands.

2.6.8 Advantages of solid fuels over the liquid fuels

i. Explosions are possible when using liquid fuels.
ii. Liquid fuels are more expensive than solid fuels.
iii. Liquid fuels can emit unpleasant odours when burning.
iv. Liquid fuels require specialised burners to burn.
v. Liquid fuels are problematic in cold climates because the oil stored in tanks must be heated to prevent oil flow from stopping.

References

[1] IEA (2023), Energy Efficiency 2023, IEA, Paris https://www.iea.org/reports/energy-efficiency-2023, Licence: CC BY 4.0. (Accessed: 08 May 2024)

[2] IEA (2023), World Energy Outlook 2023, IEA, Paris https://www.iea.org/reports/world-energy-outlook-2023, Licence: CC BY 4.0 (report); CC BY NC SA 4.0 (Annex A). (Accessed: 08 May 2024)

[3] Barney L. Cape hart, Wayne C. Turner (2016), Guide to Energy Management, Fairmont Press, ISBN: 1420084895

[4] Clive Beggs (2002), Energy: Management, Supply and Conservation, Butterworth Heinemann, ISBN: 0750650966

[5] Wayne C. Turner & Steve Doty (2012), Energy Management Handbook, 7th Ed., Fairmont Press, ISBN: 142008870X

[6] Paul W. O'Callaghan (1993), Energy Management, McGraw-Hill Professional, ISBN: 0077076788.

[7] Mehta, V.K. & Rohit Mehta (2006) Principles of power system. New Delhi: S. Chand & Company.

3

Energy Efficiency Fundamentals

Chapter Preview

This chapter reviews energy efficiency fundamentals, how we can reduce energy waste and costs. It also delves deep into energy-efficient electrical services such as power factor, variable speed drives, and efficient lighting

3.1 Energy Efficiency

Energy efficiency refers to using less energy to complete the same task or achieve the same result. Energy efficiency, as opposed to energy conservation, which focuses on using less energy overall, aims to optimise energy use. This concept is fundamental in a variety of sectors, including residential, commercial, and industrial settings, and has the potential to significantly reduce energy consumption and improve performance.

3.1.1 Reducing energy waste and costs

Energy efficiency is critical for reducing energy waste and associated costs.

1. Practical applications: Examples of how to achieve efficiency include installing energy-efficient appliances, upgrading insulation, and optimising industrial processes.
2. Economic benefits: Energy efficiency saves money for businesses and homeowners by lowering utility bills and operational costs.
3. Waste reduction: Efficient energy use reduces energy waste, which contributes to overall resource conservation.

Energy efficiency for environmental sustainability

The environmental impact of energy efficiency is profound:

1. Emission reduction: Consuming less energy allows us to significantly reduce greenhouse gas emissions and other pollutants.

2. Sustainable practices: Energy-efficient technologies and practices promote sustainable resource management, which is critical for environmental preservation.
3. Climate change mitigation: Global efforts to increase energy efficiency are critical for combating climate change and protecting ecosystems.

Energy security and global implications

Energy efficiency improves energy security and has wide-ranging global implications:

1. Strengthening energy independence: By lowering overall energy demand, countries can reduce their reliance on external energy sources.
2. Efficient energy use can help to stabilise energy markets and prices.
3. Economic and geopolitical implications: Increased energy efficiency has the potential to boost economic growth while also influencing global energy policies and dynamics.

Challenges in achieving energy efficiency

Realising energy efficiency presents challenges:

1. Initial investment and technological barriers: High upfront costs and limited access to technology can make it difficult to implement energy-efficient solutions.
2. Awareness and policy support: A lack of awareness, as well as the need for supportive policies and incentives are major factors in promoting energy efficiency.
3. Overcoming obstacles: Strategies such as government incentives, public education, and technological innovation are critical for overcoming these barriers.

3.1.2 Energy efficiency and its significance

Understanding energy efficiency

Energy efficiency is an important concept in our quest for sustainable living. It focuses on optimising energy use to achieve the same or better results while using less energy. Unlike energy conservation, which aims to reduce overall energy consumption, energy efficiency seeks to maximise output with minimal input.

Key points:

Reducing energy waste and costs: Energy efficiency is critical in combating wasteful energy practices. By using energy more efficiently, we reduce unnecessary losses and operational costs.

Differentiating energy efficiency and conservation:

- Energy efficiency: This strategy seeks to improve the performance of energy-consuming systems (such as appliances, buildings, and vehicles) by reducing energy losses. It is all about making the most of limited resources.
- Energy conservation entails reducing overall energy consumption through lifestyle changes, turning off lights, and implementing energy-saving practices.

Benefits of energy efficiency:

- Cost savings: Efficient energy use leads to lower utility bills for individuals, businesses, and industries.
- Environmental sustainability: Consuming less energy reduces the strain on natural resources and ecosystems. This contributes to a healthier environment.
- Greenhouse gas emission reduction: Energy production is a significant source of greenhouse gas emissions. We can reduce the impact of climate change by improving efficiency.
- Enhanced energy security: Efficient energy practices reduce our reliance on fossil fuels while increasing our energy resilience.

3.2 Principles of Energy Conservation

Energy conservation is an important aspect of modern energy management, which includes practices for reducing energy consumption. At its core is the fundamental law of energy conservation, which states that energy can only be transformed, not created or destroyed. This section looks at how this principle is used in a variety of professional fields to improve energy efficiency.

Understanding the law of conservation of energy

The conservation of energy is a fundamental principle of physics:

- Energy transformation: It asserts that energy in a closed system remains constant and only changes form, such as from kinetic to thermal energy.

- Implications of energy use: This principle emphasises the importance of efficient energy use and transformation, as inefficient processes result in unnecessary energy waste.

Practical applications in various fields.

Energy conservation principles have practical applications in a variety of fields.

- Building and architecture: Using energy-saving designs such as passive solar heating, effective insulation, and efficient HVAC systems.
- Industrial processes: Increasing efficiency by optimising machinery use, reducing energy waste, and recycling energy whenever possible.
- Transportation: Promoting fuel-efficient vehicles and alternative modes of transportation in order to reduce energy consumption.

Small changes, big impact

Even minor changes in energy habits can yield significant savings:

- Everyday energy conservation practices: Simple actions include turning off lights when not in use, purchasing energy-efficient appliances, and performing routine maintenance on equipment.
- Cumulative impact: When widely implemented, these small changes can significantly reduce overall energy demand and contribute to environmental conservation.

Behavioural aspects of energy conservation

Energy conservation relies heavily on behaviour.

- Encouraging energy-saving habits: Promoting habits such as responsible energy use at home and work.
- Promoting behavioural change: Strategies include education, awareness campaigns, and incentives for energy-efficient practices.

Energy conservation: Principles and practices

1. **Law of conservation of energy**:
    - The law of energy conservation is fundamental to physics. It states that energy cannot be created or destroyed; it can only be converted from one form to another. This concept is critical to understanding energy conservation.
    - When we talk about energy conservation, we mean minimising wasteful energy consumption and increasing the efficiency of energy transformations.

2. **Practical techniques for energy conservation**:

Efficient appliances and equipment:

- Choosing energy-efficient appliances and equipment has a significant impact on overall energy consumption. Look for devices with the ENERGY STAR label, which indicates that they meet stringent energy-efficiency standards.
- LED lightbulbs, energy-efficient HVAC systems, and smart thermostats are some examples.

Insulation and sealing:

- Proper building insulation and sealing prevent winter heat loss and summer heat gain.
- To save money on heating and cooling, insulate your walls, roofs, and windows and seal any gaps.

Behavioural changes:

- Encourage occupants to adopt energy-saving habits. Simple actions such as turning off lights when not in use, unplugging chargers, and adjusting thermostat settings can have a significant impact.

Renewable energy sources:

- The transition to renewable energy sources (such as solar panels, wind turbines, and geothermal systems) reduces reliance on fossil fuels.
- These sources generate clean energy and help to ensure a more sustainable future.

Industries and businesses can improve their processes by implementing energy-efficient technologies.

Energy audits, variable speed drives, and optimised production schedules are all effective strategies.

Transportation efficiency:

- Encourage public transportation, carpooling, and electric vehicles.
- Efficient transportation lowers fuel consumption and greenhouse gas emissions.

Building design and orientation:

- Architects and builders can design energy-efficient structures that take into account natural lighting, passive solar heating, and proper ventilation.

- Buildings can be oriented to maximise sunlight exposure, reducing the need for artificial lighting and heating.

Smart grids and energy management Systems:

- Smart grids allow for better monitoring and control over energy distribution.
- Energy management systems assist in tracking consumption patterns and optimising energy usage.

Education and awareness:

- Inform individuals, communities, and organisations about the value of energy conservation.
- Awareness campaigns can encourage behavioural changes and promote responsible energy consumption.

3.3 Energy Management

Effective energy management is critical for organisations that want to improve energy efficiency and sustainability. It encompasses a systematic approach to controlling and optimising energy consumption, ranging from simple behavioural changes to the integration of advanced technologies. The strategies used not only aim to reduce energy consumption, but also to improve overall operational efficiency.

Behavioural changes for energy efficiency

Behavioural change is an essential component of energy management:

- Cultivating energy-conscious habits: Encourages practices such as turning off lights and equipment when not in use and properly adjusting thermostats.
- Employee engagement and training: Getting employees involved in energy-saving initiatives and providing training to foster an energy-aware and responsible culture.

Technological implementations for energy management

Leveraging technology can significantly enhance energy management efforts:

- Energy-efficient appliances: Upgrading to appliances that use less energy while maintaining performance.
- Smart systems: Using real-time data to optimise energy consumption through the use of smart thermostats and building management systems.

- IoT devices: Using Internet of things devices to monitor and control energy consumption remotely and efficiently.

Data-driven energy management

Data is critical for making informed energy management decisions.

- Monitoring and tracking entails using systems to continuously monitor and track energy consumption across multiple areas and systems.
- Energy Audits: Conducting regular energy audits to evaluate performance and identify opportunities for improvement.
- Analytics: Analysing data to better understand consumption patterns and develop targeted energy-saving strategies.

Optimising operations and maintenance

Operational efficiency is essential for energy conservation.

- Maintenance scheduling entails keeping equipment in good working order on a regular basis.
- Equipment upgrades involve replacing or upgrading outdated or inefficient machinery with more energy-efficient models.
- Implement a preventive maintenance schedule to reduce downtime and energy waste.

Benefits of effective energy management

The advantages of implementing energy management strategies are numerous:

- Reduced operating costs: Significant cost savings achieved through reduced energy consumption and increased efficiency.
- Improved environmental performance: A smaller carbon footprint and less environmental impact.
- Regulatory compliance: Achieving or exceeding energy-related regulatory standards and requirements.
- Improving corporate reputation and competitiveness in the market.

3.3.1 Energy management strategies

1. **Behavioural changes**:
    - Encouraging employees and stakeholders to adopt energy-saving practices has a significant impact on energy efficiency. Simple actions such as turning off lights when not in use, adjusting thermostat settings, and shutting down equipment during non-working hours all help to conserve energy.

2. **Operational optimisation**:
    - Regular maintenance and calibration of machinery and equipment ensures peak performance. Properly functioning systems use less energy and reduce wastage.
    - Implementing energy-efficient practices in daily operations, such as load scheduling, can aid in the effective balance of energy supply and demand.
3. **Energy audits and monitoring**:
    - Conducting energy audits identifies areas for improvement. Auditors evaluate energy consumption patterns, identify inefficiencies, and suggest corrective actions.
    - Real-time monitoring systems track energy usage, allowing businesses to make informed decisions and respond quickly to anomalies.
4. **Technology integration**:
   Using advanced technologies increases energy efficiency. Examples include:
    - Smart sensors and controls: These automate lighting, HVAC, and other systems based on occupancy and environmental factors.
    - Energy management software offers data analytics, predictive modelling, and optimisation tools.
    - Renewable energy sources: Installing solar panels, wind turbines, or geothermal systems reduces your reliance on traditional energy sources.
5. **Employee training and awareness**:
    - Educating employees on energy-saving practices helps to foster a sustainable culture. Training programmes can address issues such as energy conservation, waste reduction, and responsible resource usage.

Benefits of effective energy management

1. **Cost savings**:
    - Reduced energy consumption translates directly into lower utility bills. Efficient energy management lowers operational costs, freeing up funds for other investments.

- Energy-efficient practices also increase the lifespan of equipment, lowering replacement costs.

2. **Environmental impact**:
   - Reduced energy consumption results in lower greenhouse gas emissions. Organisations that prioritise sustainability contribute positively to environmental preservation.
   - Implementing clean energy solutions further reduces environmental impact.
3. **Regulatory compliance**:
   - Many countries have energy-efficiency regulations and standards. Complying with these requirements ensures legal compliance and prevents penalties.
   - Organisations that manage energy proactively are better prepared to meet changing regulatory requirements.
4. **Enhanced reputation and stakeholder relations**:
   - Demonstrating a commitment to energy efficiency improves an organisation's reputation. Customers, investors, and partners value environmentally friendly practices.
   - Positive public perception can result in more business opportunities and partnerships.

3.4 Energy-efficient Electrical Services

Buildings waste a lot of energy due to poor electrical design and maintenance. The wasted energy is of the worst kind: expensive electrical energy, which can cost up to five times as much as the unit cost of heat.

Unfortunately, misguided building designers frequently overlook excessive electrical energy consumption in favour of thermal energy consumption, which is less expensive. Much energy is wasted due to poorly designed lighting, pumps, fans, and heating systems.

There are several relatively simple technologies that can be used to significantly reduce energy costs in motor drives and luminaire installations.

3.4.1 Power factor

Reactive (or inductive) electrical loads include electric induction motors and fluorescent lamp fittings. Reactive (AC circuit) electrical loads are significant because, unlike resistive loads such as incandescent lighting, they cause the

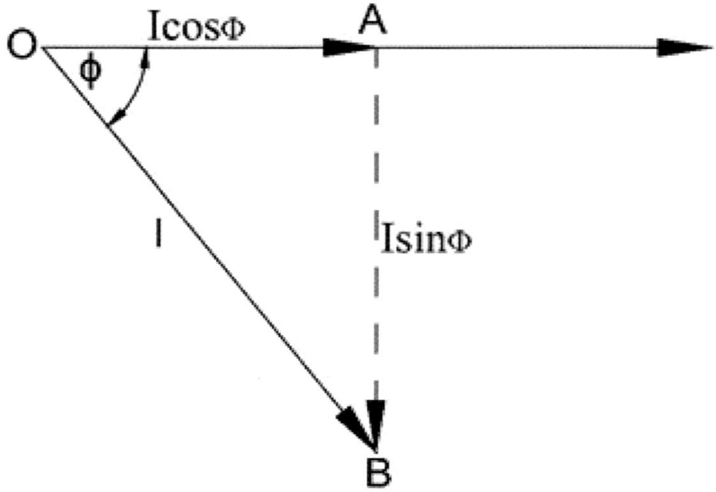

Figure 3.1 AC circuit phasor diagram.

current to become out of phase with the voltage. This simply means that inductive equipment draws more current than expected based on its useful power rating. Ultimately, the consumer is responsible for paying for this additional current.

Power factor is defined as the 'cosine of the angle between voltage and current in an alternating current circuit'.

In an alternating current (AC) circuit, voltage and current have a phase difference (φ). The term 'cos φ' refers to the circuit's power factor. Consider an inductive circuit that draws a lagging current I from the supply voltage V, with the angle of lag being φ. Figure 3.1 depicts the phasor diagram of the circuit.

- If the circuit is inductive, the current lags behind the voltage and the power factor is referred to as lagging.
- However, in a capacitive circuit, current leads to the voltage and power factor, which are said to be leading.

The component $I \cos\varphi$ is known as the active or wattful component, while component $I \sin\varphi$ is called the reactive or wattless component. The reactive component is a measure of the power factor. If the reactive component is small, the phase angle φ is small, and hence, the power factor cos φ will be high. Therefore, a circuit with a low reactive current (i.e. $I \sin\varphi$) will have

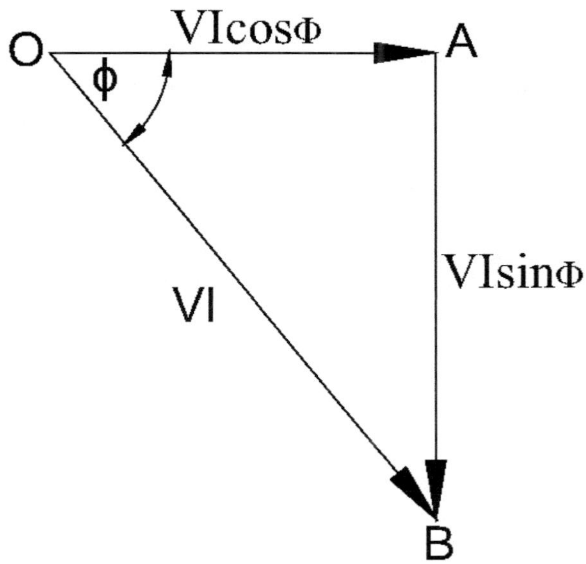

Figure 3.2 Power factor triangle.

a high power factor and *vice-versa*. It should be noted that the value of the power factor can never be more than unity.

- It is a usual practice to attach the word 'lagging' or 'leading' with the numerical value of the power factor to signify whether the current lags or leads the voltage. Thus, if the circuit has a p.f. of 0.5 and the current lags the voltage, we generally write p.f. as 0.5 lagging.
- Sometimes, the power factor is expressed as a percentage. Thus, 0.8 lagging power factor may be expressed as 80% lagging.

3.4.1.1 Power triangle

- $OA = VI\cos\varphi$ and represents the *active power* in watts or kW
  $AB = VI\sin\varphi$ and represents the *reactive power* in VAR or kVAR.
  $OB = VI$ and represents the *apparent power* in VA or kVA.

$$\text{Power factor, } \cos\theta = \frac{OA}{OB} = \frac{\text{Active power}}{\text{apparent power}} = \frac{kW}{kVA}. \tag{3.1}$$

Thus, the power factor of a circuit can be defined as the ratio of active power to apparent power.

Lagging* reactive power is to blame for the low power factor. The power triangle clearly shows that the smaller the reactive power component, the higher the circuit power factor. However, when the circuit current exceeds the voltage, the reactive power is referred to as leading reactive power.

For leading currents, the power triangle is reversed. This fact provides a key to improving power factor. If a device that consumes leading reactive power (for example, a capacitor) is connected in parallel with the load, the load's lagging reactive power is partially neutralised, improving the load's power factor.

The power factor of a circuit can be defined in one of the following three ways (Figure 3.2):

$$\begin{aligned}\text{Power factor} &= \cos\theta = \text{cosine of angle between } V \text{ and } I\\ \text{Power factor} &= \frac{R}{Z} = \frac{\text{Resistance}}{\text{Impedance}}\\ \text{Power factor} &= \frac{VI\cos\theta}{VI} = \frac{\text{Active power}}{\text{Apparent power}}.\end{aligned} \tag{3.2}$$

The reactive power is neither consumed in the circuit nor does it do any useful work. It merely flows back and forth in both directions in the circuit. A watt meter does not measure reactive power.

Suppose a circuit draws a current of 10 A at a voltage of 200 V, and its p.f. is 0.8 lagging:

$$\begin{aligned}\text{Apparent power} &= VI = 200 \times 10 = 2000 \text{ VA}\\ \text{Active power} &= VI\cos\theta = 200 \times 10 \times 0.8 = 1600 \text{ W}\\ \text{Reactive power} &= VI\sin\theta = 200 \times 10 \times 0.6 = 1200 \text{ VAR}\end{aligned}$$

The circuit receives an apparent power of 2000 VA and is able to convert only 1600 watts into active power. The reactive power is 1200 VAR and does no useful work. It merely flows into and out of the circuit periodically. In fact, reactive power is a liability to the source because the source has to supply the additional current (i.e. $I\sin\varphi$).

3.4.1.2 Disadvantages of low power factor

The power factor plays an essential role in a.c. circuits since the power consumed depends on this factor.

For single-phase supply,

$$P = V_L I_L \cos\theta, \tag{3.3}$$

$$I_L = \frac{P}{V_L \cos\theta}. \tag{3.4}$$

For three-phase supply,

$$P = \sqrt{3} V_L I_L \cos\theta, \tag{3.5}$$

$$I_L = \frac{P}{\sqrt{3} V_L \cos\theta}. \tag{3.6}$$

It is clear from above that for fixed power and voltage, the load current is inversely proportional to the power factor. The lower the power factor, the higher is the load current and *vice-versa*. A power factor less than unity results in the following disadvantages:

(i) **Large kVA rating of equipment.** The electrical machinery (e.g. alternators, transformers, switchgear) is always rated in *kVA.

$$\text{Say, } kVA = \frac{kW}{\cos\theta}. \tag{3.7}$$

**The electrical machinery is rated in kVA because the power factor of the load is not known when the machinery is manufactured in the factory.*

It is clear that the kVA rating of the equipment is inversely proportional to the power factor. The smaller the power factor, the larger is the kVA rating. Therefore, at a low power factor, the kVA rating of the equipment has to be made higher, making the equipment larger and more expensive.

(ii) **Greater conductor size.** To transmit or distribute a fixed amount of power at constant voltage, the conductor will have to carry more current at a low power factor. This necessitates a large conductor size. For example, take the case of a single phase a.c. motor having an input of 10 kW on full load, the terminal voltage being 250 V. At unity p.f., the input full-load current would be 10,000/250 = 40 A. At 0.8 p.f; the kVA input would be 10/0.8 = 12.5 and the current input 12,500/250 = 50 A. If the motor is worked at a low power factor of 0.8, the cross-sectional area of the supply cables and motor conductors would have to be based upon a current of 50 A instead of 40 A, which would be required at unity power factor.

(iii) **Large copper losses.** The large current at low power factor causes more I^2R losses in all the elements of the supply system. This results in poor efficiency.

(iv) **Poor voltage regulation.** The large current at a low lagging power factor causes greater voltage drops in alternators, transformers, transmission lines, and distributors. This results in the decreased voltage available at the supply end, thus impairing the performance of utilisation devices. In order to keep the receiving end voltage within permissible limits, extra equipment (i.e. voltage regulators) is required.

(v) **Reduced handling capacity of the system.** The lagging power factor reduces the handling capacity of all the elements of the system. This is because the reactive component of the current prevents the full utilisation of installed capacity.

3.4.1.3 Causes of low power factor

A low power factor is undesirable from an economic point of view. Normally, the power factor of the whole load on the supply system is lower than 0.8. The following are the causes of low power factor:

(i) Most of the a.c. motors are of induction type (1φ and 3φ induction motors) which have a low lagging power factor. These motors work at a power factor which is extremely small on light load (0.2–0.3) and rises to 0.8 or 0.9 at full load.

(ii) Arc lamps, electric discharge lamps, and industrial heating furnaces operate at a low lagging power factor.

(iii) The load on the power system is varying, being high during morning and evening and low at other times. During the low load period, the supply voltage increases, which increases the magnetisation current. This results in a decreased power factor.

3.4.1.4 Power factor improvement

The low power factor is primarily due to the fact that the majority of power loads are inductive and thus require lagging currents. To improve the power factor, some devices that use leading power should be connected in parallel with the load. Capacitors are one example of such devices. The capacitor draws a leading current, which partially or completely neutralises the load current's lagging reactive component. This improves the power factor of the load.

3.4.1.5 Power factor improvement equipment

Normally, the power factor of the whole load on a large generating station is in the region of 0.8–0.9. However, sometimes, it is lower, and in such cases, it is generally desirable to take special steps to improve the power factor. This can be achieved by the following equipment.

(i) Static capacitors
(ii) Synchronous condenser
(iii) Phase advancers

3.4.1.6 Calculations of power factor correction

Consider an inductive load taking a lagging current I at a power factor $\cos\varphi_1$. In order to improve the power factor of this circuit, the remedy is to connect such equipment in parallel with the load, which takes a leading reactive component and partly cancels the lagging reactive component of the load. Figure 3.3 (*i*) shows a capacitor connected across the load. The capacitor takes a current I_C which leads the supply voltage V by 90°. The current I_C partly cancels the lagging reactive component of the load current as shown in the phasor diagram in Figure 3.3 (*ii*). The resultant circuit current becomes I'

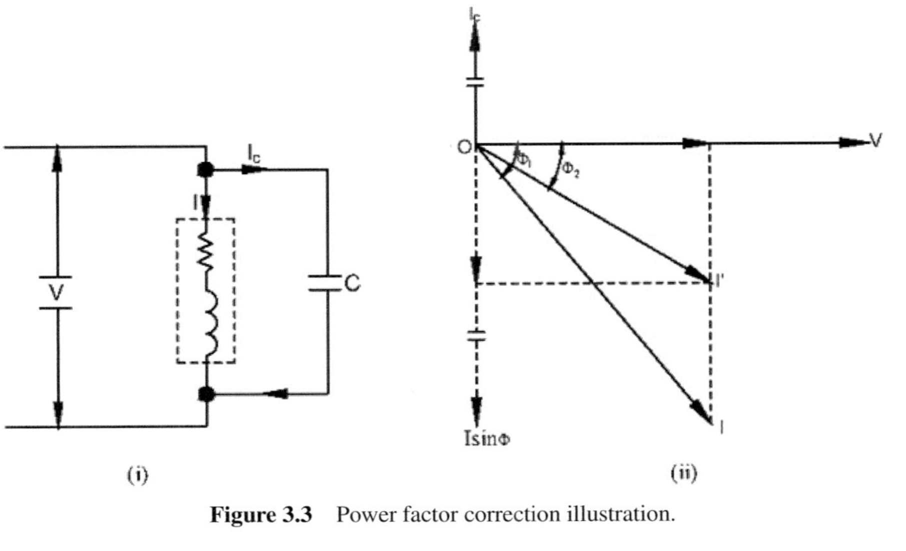

Figure 3.3 Power factor correction illustration.

and its angle of lag is φ_2. It is clear that φ_2 is less than φ_1 so that new p.f. cosφ_2 is more than the previous p.f. cosφ_1.

From the phasor diagram, it is clear that after p.f. correction, the lagging reactive component of the load is reduced to I' sinφ_2.

$$\begin{aligned} I' \sin\phi_2 &= I \sin\phi_1 - I_C \\ I_C &= I \sin\phi_1 - I' \sin\phi_2. \end{aligned} \tag{3.8}$$

The capacitance of the capacitor to improve p.f. from cosφ_1 to cosφ_2

$$= \frac{I_C}{wV} \left(\text{Because, } X_C = \frac{V}{I_C} = \frac{1}{wC} \right)$$

$$\boldsymbol{X_C = \frac{V}{I_C} = \frac{1}{wC}}. \tag{3.9}$$

Power triangle

The power factor correction can also be illustrated from the power triangle. Thus, referring to Figure 3.4, the power triangle *OAB* is for the power factor cosφ_1, whereas the power triangle *OAC* is for the improved power factor cosφ2. It may be seen that active power (*OA*) does not change with power factor improvement. However, the lagging kVAR of the load is reduced by the p.f. correction equipment, thus improving the p.f. to cosφ_2

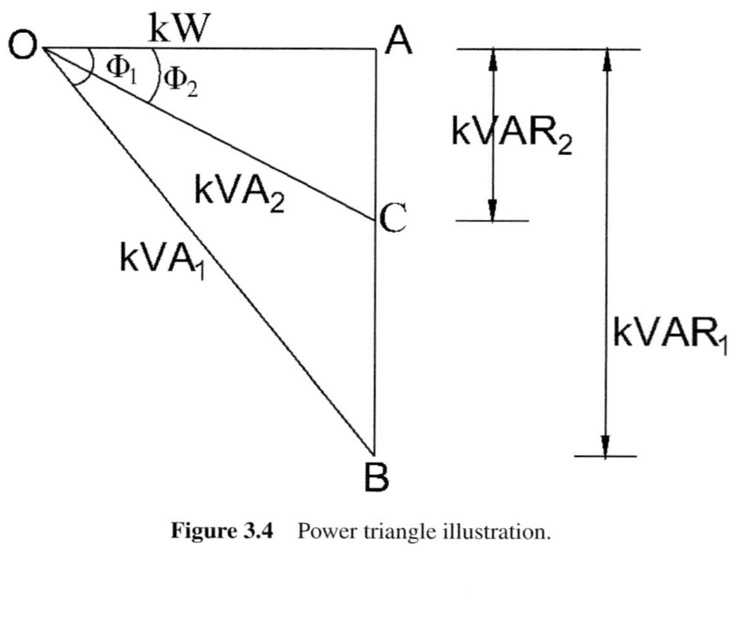

Figure 3.4 Power triangle illustration.

Leading kVAR supplied by p.f. correction equipment

$$
\begin{aligned}
&= BC = AB - AC \\
&= \mathrm{kVAR}_1 - \mathrm{kVAR}_2 \\
&= OA(\tan\phi_1 - \tan\phi_2) \\
&= kW(\tan\phi_1 - \tan\phi_2)
\end{aligned}
$$

Knowing the leading kVAR supplied by the p.f. correction equipment, the desired results can be obtained.

Example 3.1

An alternator is supplying a load of 300 kW at a p.f. of 0.6 lagging. If the power factor is raised to unity, how many more kilowatts can the alternator supply for the same kVA loading?

Solution:

$$
\begin{aligned}
\mathrm{kVA} &= \frac{kW}{\cos\phi} = \frac{300}{0.6} = 500\ \mathrm{kVA} \\
kW \text{ at } 0.6 \text{ p.f.} &= 300\ \mathrm{kW} \\
kW \text{ at } 1 \text{ p.f.} &= 500 \times 1 = 500\ \mathrm{kW}
\end{aligned}
$$

∴ Increased power supplied by the alternator = 500 – 300 = **200 kW**

Take note of the significance of power factor improvement. When the alternator's p.f. is unity, 500 kVA equals 500 kW, and the engine driving the alternator must be capable of developing this power while accounting for the alternator's losses. However, when the load has a power factor of 0.6, the power output is only 300 kW. As a result, the engine generates only 300 kW, despite the alternator's rated output of 500 kVA.

Example 3.2

A single-phase motor connected to 400 V, 50 Hz supply takes 31.7A at a power factor of 0.7 lagging. Calculate the capacitance required in parallel with the motor to raise the power factor to 0.9 lagging.

Solution:

The circuit and phasor diagrams are shown in Figures 3.5 and 3.6, respectively. Here, motor M is taking a current I_M of 31.7 A. The current I_C taken by the capacitor must be such that when combined with I_M, the resultant current I lags the voltage by an angle φ where $\cos\varphi = 0.9$.

Referring to the phasor diagram in Figure 3.6,

$$
\text{Active component of } I_M = I_M \cos\phi_M = 31.7 \times 0.7 = 22.19\ \mathrm{A}
$$

$$\text{Active component of } I = I\cos\phi = I \times 0.9$$

These components are represented by *OA* in Figure 3.6

$$I = \frac{22.19}{0.9} = 24.65 \text{ A}$$

$$\text{Reactive component of } I_M = I_M \sin\phi_M = I_M\sqrt{1-\cos^2\phi_M}$$

$$= 31.7 \times \sqrt{1-(0.7)^2} = 31.7 \times 0.714 = 22.6 \text{ A}$$

$$\text{Reactive component of } I = I\sin\phi = 24.65\sqrt{1-(0.9)^2}$$

$$= 24.65 \times 0.436 = 10.75 \text{ A}$$

$$I_C = \text{Reactive component of } I_M - \text{Reactive component of } I$$

$$= 22.6 - 10.75 = 11.85 \text{ A}$$

$$But,\ I_C = \frac{V}{X_C} = V \times 2\pi f C$$

$$11.85 = 400 \times 2\pi \times 50 \times C$$

$$C = 94.3 \times 10^{-6} \text{ F} = 94.3\ \mu\text{F}$$

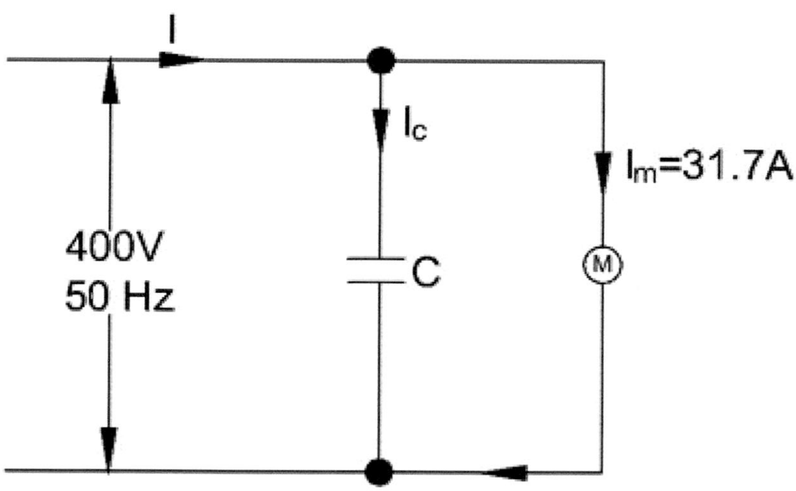

Figure 3.5 Circuit diagram for example 3.2.

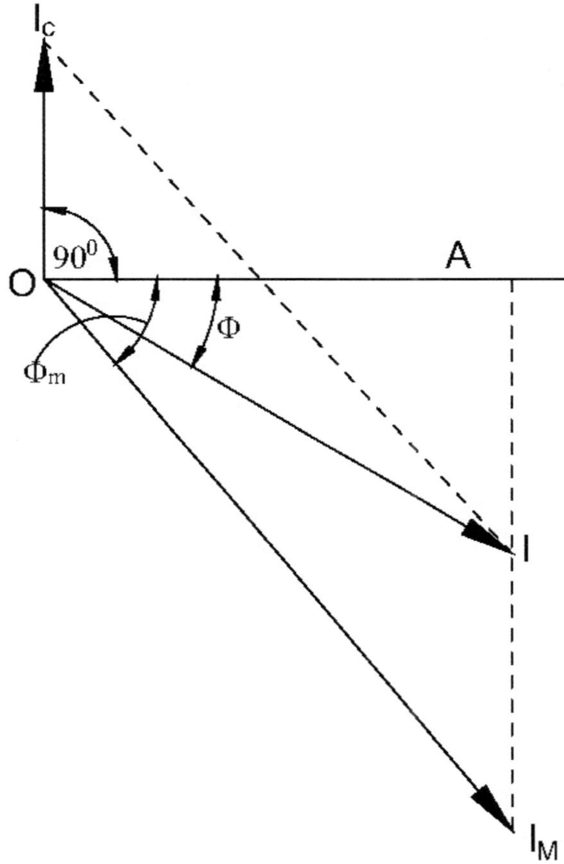

Figure 3.6 Phasor diagram for example 3.2.

3.4.1.7 Importance of power factor improvement

The improvement of power factor is significant for both consumers and generating stations, as discussed below:

- For consumers. A consumer must pay electricity charges based on his maximum demand (kVA) plus the units consumed. If the consumer's power factor improves, his maximum kVA demand decreases, resulting in an annual savings due to maximum demand charges. Although power factor improvement requires additional annual expenditure for p.f. correction equipment, it results in a net annual savings for the consumer.

- For generating stations. A generating station is just as concerned with power factor improvement as the consumer. A power station's generators are rated in *kVA*, but their useful output is based on kW output. The power factor determines the number of units supplied by the station, as its output is kW $=$ kVA $\times$ cosφ. The higher the power factor of the generating station, the more kWh it provides to the system. This leads us to the conclusion that an improved power factor increases the power station's earning capacity.

3.4.1.8 Most economical power factor

If a consumer's power factor improves, his maximum kVA demand decreases, resulting in annual savings over maximum demand charges. However, improving the power factor requires a capital investment in power factor correction equipment. Every year, the consumer will pay annual interest and depreciation on the investment in p.f. correction equipment. As a result, the net annual savings will be the difference between the annual savings in maximum demand charges and the annual expenditure on p.f. correction equipment.

The value to which the power factor should be improved so as to have maximum net annual saving is known as the most economical power factor.

Consider a consumer taking a peak load of *P* kW at a power factor of cosφ_1 and charged at a rate of \$*x* per kVA of maximum demand per annum.

Suppose the consumer improves the power factor to cosφ_2 by installing p.f. correction equipment. Let expenditure incurred on the p.f. correction equipment be \$*y* per kVAR per annum. The power triangle at the original p.f. cosφ_1 is OAB, and for the improved p.f. cosφ_2, it is OAC (See Figure 3.7.)

$$\text{kVA max.demand at } \cos\phi_1, \text{kVA}_1 = P/\cos\phi_1 = P\sec\phi_1$$

$$\text{kVA max.demand at } \cos\phi_2, \text{kVA}_2 = P/\cos\phi_2 = P\sec\phi_2$$

Annual savings in maximum demand charges

$$\begin{aligned}
&\text{Ksh.}x(\text{kVA}_1 - \text{kVA}_2)\\
&\text{Ksh.}x(P\sec\phi_1 - P\sec\phi_2)\\
&= \text{Ksh.}xP\,(\sec\phi_1 - \sec\phi_2) \ldots\ldots\ldots \qquad (3.10)\\
&\text{Reactive power at } \cos\phi_1, \text{kVAR}_1 = P\tan\phi_1\\
&\text{Reactive power at } \cos\phi_2, \text{kVAR}_2 = P\tan\phi_2
\end{aligned}$$

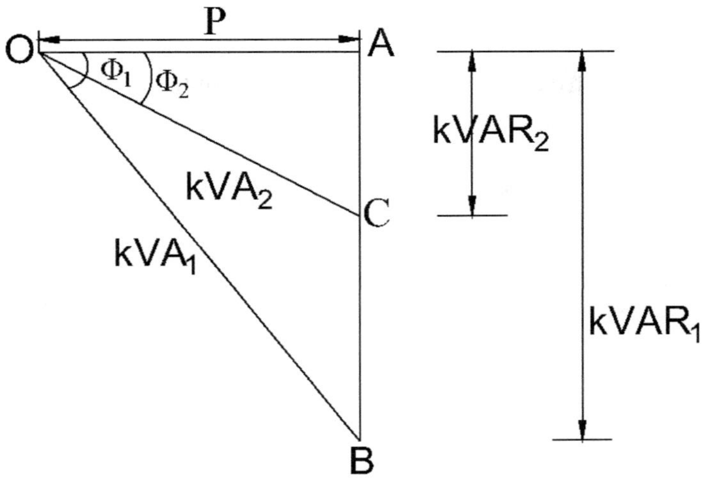

Figure 3.7 Most economical power illustration.

Leading kVAR taken by p.f. correction equipment

$$= P(\tan\phi_1 - \tan\phi_2)$$

Annual cost of p.f. correction equipment

$$= \text{Ksh.}Py\,(\tan\kappa\phi_1 - \tan\phi_2)\ldots\ldots\ldots\ldots \quad (3.11)$$

Net annual saving, S = eqn (3.10) – eqn (3.11)

$$= xP\,(\sec\phi_1 - \sec\phi_2) - yP(\tan\phi_1 - \tan\phi_2).$$

In this expression, only φ_2 is variable, while all other quantities are fixed. Therefore, the net annual saving will be maximum if differentiation of the above expression *w.r.t.* φ_2 is zero, i.e.

$$\frac{dy}{d\phi_2}\,(S) = 0$$

$$\frac{dy}{d\phi_2} = [xP\,(\sec\phi_1 - \sec\phi_2) - yP(\tan\phi_1 - \tan\phi_2)] = 0$$

$$\frac{dy}{d\phi_2}\,(xP\sec\phi_1) - \frac{dy}{d\phi_2}\,(xP\sec\phi_2) - \frac{dy}{d\phi_2}\,(yP\tan\phi_1)$$

$$+yP\frac{dy}{d\phi_2}\,(\tan\phi_2) = 0$$

$$0 - xP\sec\phi_2\tan\tan\phi_2 - 0 + yP\sec^2\phi_2 = 0$$

$$\begin{aligned} -x\tan\phi_2 + y\sec\phi_2 &= 0 \\ \tan\phi_2 &= \frac{y}{x}\sec\phi_2 \\ \sin\phi_2 &= y/x \\ \text{Most economical power factor, } \cos\phi_2 &= \sqrt{1-\sin^2\phi_2} = \sqrt{1-(y/x)^2} \end{aligned}$$

It may be noted that the most economical power factor ($\cos\varphi_2$) depends upon the relative costs of supply and p.f. correction equipment but is independent of the original p.f. $\cos\varphi_1$.

Example 3.3

A factory which has a maximum demand of 175 kW at a power factor of 0.75 lagging is charged at $72 per kVA per annum. If the phase-advancing equipment costs $120 per kVAR, find the most economical power factor at which the factory should operate. Interest and depreciation total 10% of the capital investment on the phase-advancing equipment.

$$\begin{aligned} &\text{Power factor of the factory, } \cos\phi_1 = 0.75 \text{ lagging} \\ &\text{Max.demand charges, } x = \text{Ksh.72 per kVA annum.} \\ &\text{Expenditure on phase advancing equipment,} \\ &\qquad y = \text{Ksh.}120 \times 0.1 = \text{Ksh.}\frac{\frac{12}{\text{kVAR}}}{\text{annum}}. \end{aligned}$$

The total investment for producing 1 kVAR is $120. The annual interest and depreciation is 10%. It means that an expenditure of $120 × 10/100 = $12 is incurred on 1 kVAR per annum.

The most economical p.f. at which the factory should operate is

$$\cos\phi_2 = \sqrt{1-\left(\frac{y}{x}\right)^2} = \sqrt{1-(12/72)^2} = 0.986.$$

3.4.2 Electric motors

Induction motors are commonly used in a variety of applications. Pumps, fans, compressors, escalators, and lifts are all powered by various types of motors. Most modern buildings rely on induction motors to operate. Furthermore, electric motors are frequently the most expensive items for plants to operate in many office buildings. It is therefore well worth understanding how induction motors use electrical energy and investigating potential energy potential energy-saving measures.

All induction motors have inherent inefficiencies. These energy losses include:

- Iron losses which are associated with the magnetic field created by the motor. They are voltage-related and, therefore, constant for any given motor and independent of load.
- Copper losses (or I^2R losses) which are created by the resistance of the copper wires in the motor. The greater the resistance of the coil, the more heat is generated and the greater the power loss. These losses are proportional to the square of the load current.
- Friction losses which are constant for a given speed and independent of load.

These losses can be divided into those that vary with motor load and those that are constant, whatever the load. When a motor is running at full load, the split between the two is about 70% and 30%, respectively. Under part load, this split changes; at low load, the current drawn is small, and the I^2R losses are low.

As a result, iron losses predominate, and because they are caused by reactive current consumption, the power factor is low. Even at full load, induction motors have a relatively low power factor, usually around 0.8.

3.4.2.1 Motor sizing

Correct sizing of electric motors is critical to their efficient operation since oversized motors tend to exhibit poor power factors and lower efficiencies. Depending on size and speed, a typical standard motor may have a full load efficiency between 55% and 95%. Generally, the lower the speed, the lower the efficiency and lower the power factor.

Typically, motors have efficiencies that are fairly consistent down to about 75% full load. Thereafter, they may lose approximately 5%–50% of full load, after which the efficiency rapidly declines (as shown in Figure 3.8).

The performance curve in Figure 3.8 shows that it is possible to oversize a motor by up to 25% without significantly reducing its efficiency, as long as the motor is run at a fairly constant load. If the load fluctuates and rarely reaches 75% of full load, the motor's efficiency and power factor will suffer.

Under part load conditions, the power factor typically decreases faster than the efficiency. As a result, oversized motors require more power factor correction. Oversizing a motor also raises the capital cost of the switch gear and wiring that serves it.

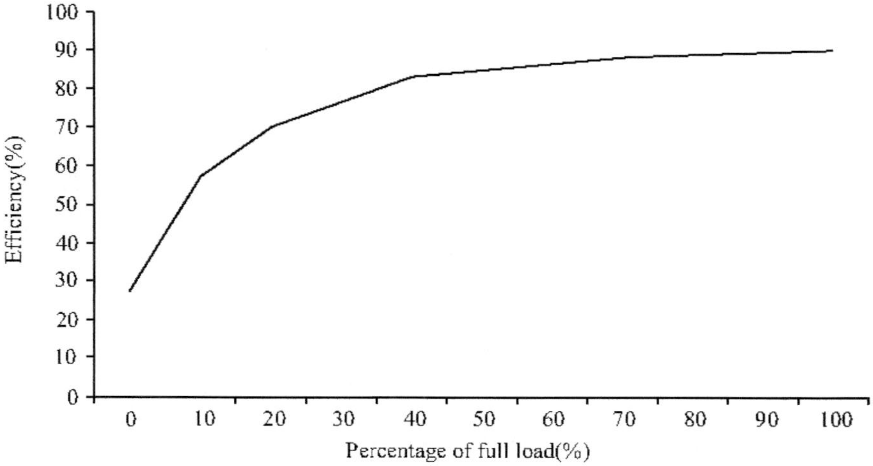

Figure 3.8 Relationship between motor loading and efficiency.

3.4.2.2 Variable speed drives (VSD)

The majority of induction motors used in buildings power fans or pumps. Traditionally, pipe work and duct work systems have been designed with oversized pumps and fans, followed by the use of commissioning valves and dampers to control flow rate by increasing system resistance. Mechanical constrictions can control fan and pump flow rates (see Figure 3.9), but they also increase system resistance and energy loss. This is an undesirable situation and one of the primary reasons for the high energy consumption associated with fans and pumps in many buildings.

To control flow rate, consider reducing the speed of the fan or pump motor instead of using valves and dampers.

Example 3.4

It is proposed to use a forward-curved centrifugal fan in a mechanical ventilation system. The fan is required to deliver a volume flow rate of 1.8 m^3/s, and the estimated system resistance is 500 Pa. However, the proposed fan delivers 2.06 m^3/s against a resistance of 500 Pa while running at a speed of 1440 rpm. Determine the fan power input if:

a) A volume control damper is used to achieve a volume flow rate of 1.8 m^3/s by increasing the total system resistance to 750 Pa.
b) The fan speed is reduced in order to deliver 1.8 m^3/s.

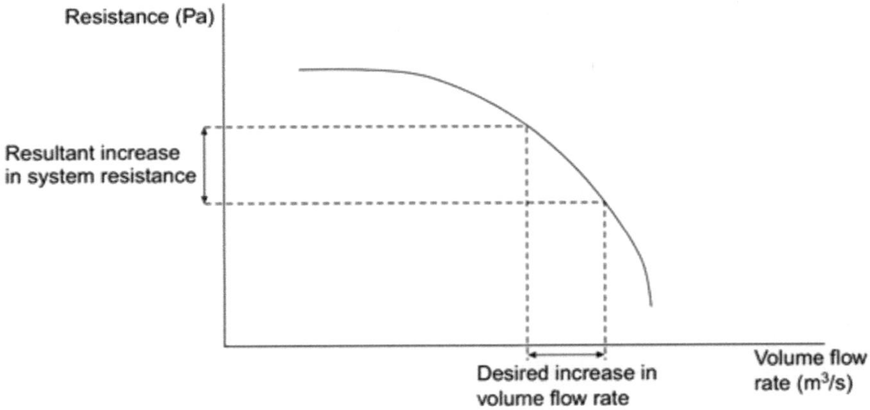

Figure 3.9 Impact of a volume control damper on system resistance.

Solution

a) Fan air power input:

$$W = V \times P_t, \tag{3.12}$$

where v is the air volume flow rate (m^3/s), and P_t is the total system resistance (Pa). Let W_1 be the fan power when delivering 2.06 m^3/s against a resistance of 500 Pa, and W_2 be the fan power when delivering 1.8 m^3/s against a resistance of 750 Pa.

$$W_1 = 2.06 \times 500 = 1030 \text{ W}$$

And

$$W_2 = 1.8 \times 750 = 1350 \text{ W}$$

Thus,

$$\text{Increase in power consumption} = \frac{1350 - 1030}{1030} \times 100 = 31.1\%.$$

b) The fan laws state that:

$$V \propto N$$

And

$$W \propto N^3,$$

where v is the air volume flow rate (m^3/s), N is the fan speed (rpm), and W is the fan air power input (W).

Let N_1 be the fan speed when delivering 2.06 m^3/s against a resistance of 500 Pa, N_3 be the fan speed when delivering 1.8 m^3/s, and W_3 be the fan power when delivering 1.8 m^3/s.

$$N_3 = 1440 \times \frac{1.8}{2.06} = 1258.3 \text{ rpm}$$

$$W_3 = 1030 \times \frac{1.8^3}{2.06^3} = 687.2 \text{ W}.$$

Therefore:
Reduction in power consumption (W_3 compared with W_1)

$$= \frac{1030 - 687}{1030} \times 100 = 33.3\%.$$

However:
Reduction in power consumption (W_3 compared with W_2)

$$= \frac{1350 - 687}{1350} \times 100 = 49.1\%.$$

It can be seen from Example 3.4 that:

- The use of volume control dampers to regulate air flow significantly increases fan energy consumption. The precise magnitude of this increase will depend on the characteristics of the particular fan selected.
- Reducing the fan speed to regulate the air flow rate always results in fan energy savings.

Reducing fan speeds results in significant fan power savings, especially when compared to the fan power increase caused by the use of volume control dampers. As a result, controlling fan and pump speeds offers significant advantages.

The energy savings shown in Example 3.4 are representative of the types of savings that can be achieved by using VSDs on fans and pumps. In most applications, using VSDs on pumps, fans, and compressors has the potential to save significant amounts of energy. Most designers overestimate system resistances, so most pumps and fans are theoretically oversized before they are selected. During the selection process, the cautious designer is unlikely to find a fan or pump that meets the theoretical 'calculated' specification, so a larger one is chosen that is certain to perform the required task. This strategy protects the system designer by preventing him or her from under sizing the fans or pumps.

Unfortunately, it also ensures that the system is significantly overdesigned, necessitating the use of volume control dampers and dampers during the commissioning process to reduce the volume flow rate. As a result, the system's capital and future operating costs have increased significantly. VSDs can ensure that even oversized fans and pumps do not significantly increase energy consumption. As a result, installing VSDs is one of the least expensive energy efficiency measures available. VSDs are expected to have payback periods of less than two years.

Using VSDs on variable volume flow systems can result in even greater energy savings compared to constant flow systems. When the load profiles and duty cycles of heating, air-conditioning, and ventilation systems are thoroughly examined, it is discovered that the majority of them consistently operate well below their intended design specifications. The main reason for this is that system designers are overly cautious during the design process. As a result, large systems with variable temperature and constant flow rate are designed. While this approach works in practice, it means that pump and fan running costs remain constant and high regardless of the operating load. Keeping the temperature constant and adjusting the flow rate can reduce pump and fan running costs as the load decreases. The variable air volume (VAV) air conditioning system is a classic example of this approach, and VSDs are well suited to it.

3.4.2.3 Principles of VSD operation

Modern electronic VSD systems control motor speed by adjusting the alternating current from the mains. Several electronic VSD systems are available. One of the most common types is the variable frequency drive, which controls speed by varying the voltage and frequency output. Such drives regulate the voltage to the motor in proportion to the output frequency, ensuring that the voltage-frequency ratio remains relatively constant. Changes in motor speed are accomplished by modulating the voltage and frequency to the motor. Figure 3.10 depicts the basic components of a variable frequency drive (VSD) system.

Variable frequency drive systems have two main components: a rectifier and an inverter. The rectifier converts standard alternating current (ac) (e.g. 240 V and 50 Hz) into adjustable direct current (DC), which is then fed into the inverter. The inverter uses electronic switches to turn DC power on and off, resulting in pulsed AC power output. This can then be controlled

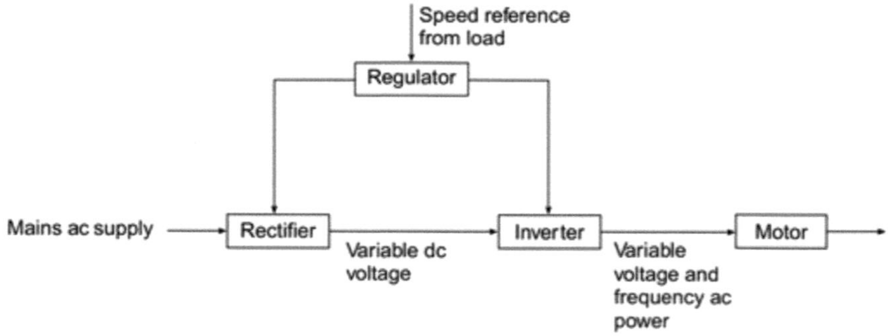

Figure 3.10 Components of a variable speed drive.

to generate the desired frequency and voltage. A regulator modifies the inverter's switching characteristics to allow for control over the output frequency.

The inverter is a critical component of a VSD system. One type of inverter that is currently in use is the pulse width modulated (PWM) inverter, which takes a fixed DC voltage from the rectifier and adjusts the output voltage and frequency. The PWM inverter generates a current waveform that closely resembles the pure sine wave of the main AC supply.

3.4.3 Checklist for electrical systems for energy Conservation

This checklist is intended to assist facilities in ensuring that their electrical systems operate efficiently, conserving energy and reducing costs.

1. Maintenance of electrical equipment

Item to check:

Inspect all electrical equipment in the building to ensure that it is well-maintained.

Corrective action:

- Establish and enforce a preventative maintenance programme.
- Identify, repair, or replace inefficient or failing equipment to increase energy efficiency.

2. Electric motors

Item to check:

Keep a detailed inventory of all large electric motors, including their locations, voltage, current, and horsepower ratings.

Corrective action:

Replace or repair motors that have problems such as incorrect voltage or current, incorrect horsepower ratings, loose connections, unbalanced voltage (for three-phase motors), poor grounding, or excessive wear on bearings or pulleys.

3. Electrical control equipment

Item to check:

- Check that all electrical control equipment is functioning properly.

Corrective action:

- Repair or replace any defective control equipment.
- Ensure that adjustments to control equipment are made only by authorised personnel.

4. Electric water heaters

Item to check:

Examine water heaters to ensure that the temperature settings are as low as possible and that heaters are not turned on when not needed.

Corrective action:

Install timers to reduce water heater operation time during periods of low or no demand.

5. Voltage drop in branch circuits

Item to check:

- Check the operating voltages of equipment located far from the power distribution panel to ensure they are not experiencing low voltage, which reduces efficiency.

Corrective action:

- Replace thinner, older wires with larger gauge wires to reduce voltage drop.
- Consider redesigning the electrical distribution system to eliminate long wire runs between distribution panels.

6. Electrical demand

Item to check:

- Compare the building's average electrical usage (kW) to peak usage (kW) to see if there is any significant difference.

Corrective action:

- If there is a significant difference, install demand controllers, load shedders, or timing devices to ensure more consistent electrical power consumption.

7. Power factor of equipment

Item to check:

Check for a power-factor penalty on the electricity bill in facilities with large electric motors, particularly in industrial settings.

Corrective action:

Ensure that motors are properly loaded during operation, as under loading can reduce the power factor.

Consider installing power-factor-correcting capacitors or a three-phase synchronous capacitor motor unit to improve power factor and lower penalties.

3.4.4 Lighting energy consumption

Electric lighting consumes a significant amount of energy in most buildings. Although the heating system consumes more energy than lighting in many buildings, the energy costs associated with lighting are frequently much higher than those associated with heating. Lighting schemes that are carefully designed and maintained can result in significant energy cost savings.

3.4.4.1 Daylighting

Daylight has the potential to significantly improve building lighting by reducing reliance on artificial lighting.

The depth of a room, the size and location of windows, the glazing system, and any external obstructions all have a significant impact on interior daylighting. These factors are typically determined during the initial design stage. Through proper early planning, it is possible to create a building that is both energy efficient and attractive on the inside. However, glazing can impose severe constraints on a building's form and operation. Poor fenestration design decisions can result in a building that is uncomfortable for the occupants and consumes a lot of energy. Glazing should therefore be handled with care. Skylights and solar Tubes (Also called 'Sun Tunnels') are also both great ways to add natural light in a building (Figure 3.11).

Figure 3.11 Solar tube and skylight.

3.4.4.2 Lighting definitions and design

A discussion of the subject of lighting design is beyond the scope of this chapter.

In brief, the performance of an artificial lighting scheme is influenced by:

- The efficacy of the lamps (i.e., the light output per watt of electrical power consumed).
- The luminaire performance.
- The layout of the luminaire fittings.
- The surface reflectance of the decor and furnishing.
- The maintenance standards.

All of these factors must be considered when designing any lighting scheme. The lumen design method is a popular approach that takes into account all of these factors. The lumen method, which allows for the quick and easy design of regular lighting schemes, is especially popular as a design method.

3.4.4.3 Energy-efficient lighting

The main factors which influence the energy consumption of lighting schemes are:

- the light output per watt of electrical power consumed (i.e., lamp efficacy);
- luminaire performance;

- the number of luminaires and their location;
- the reflectance of internal room surfaces;
- maintenance and procedure standards;
- duration of operation;
- the switching and control techniques used.

3.4.4.4 Lighting controls

The proper use of lighting controls can lead to significant energy savings. Savings are primarily achieved by maximising daylight and turning off electric lighting when not in use. To optimise energy savings, it is crucial to consider the space's occupancy pattern when designing a lighting control strategy.

There are four basic methods by which lighting installations can be controlled:

- Time-based control
- Daylight-linked control
- Occupancy-linked control
- Localised switching

Time signals may originate from local solid-state switches or building management systems. These signals turn the lights on and off at specific times. Local overrides are necessary to ensure that lighting is restored if the occupants require it.

Photoelectric cells can be used to simply turn on and off lights or to dim them. They can be mounted either externally or internally. However, time delays must be incorporated into the control system to avoid repeated rapid switching caused by fast-moving clouds. An internally mounted photoelectric dimming control system can ensure that the sum of daylight and electric lighting always reaches the design level by sensing the total light in the controlled area and adjusting the output of the electric lighting accordingly. If daylight meets design requirements, electric lighting can be turned off. The energy-saving potential of dimming control exceeds that of a simple photoelectric switching system. Dimming control is also more likely to appeal to room occupants.

Infrared, acoustic, ultrasonic, or microwave sensors can detect movement or noise in room spaces to enable occupancy-based control. These sensors activate lighting when occupancy is detected and turn it off after a predetermined time if no movement is detected. They are designed to override manual switches and prevent lighting from being left on in unoccupied spaces. This type of system requires a built-in time delay because occupants frequently remain still or quiet for short periods of time and do not appreciate being plunged into darkness if they are not constantly moving around.

Localised switching should be used in applications with large spaces. Local switches give individual occupants control over their visual environment while also promoting energy savings. Localised switching allows you to turn off artificial lighting in specific areas while keeping it on in other areas that require it, which is impossible if the lighting for an entire space is controlled by a single switch.

3.4.4.5 Maintenance

Luminaires and room surfaces become dirty over time, resulting in decreased lamp output. Lamps also fail and require replacement. As a result, all lighting installations lose performance over time. To ensure that an installation runs efficiently, regular maintenance is required. Simple cleaning of lamps and luminaires can significantly improve lighting performance.

Therefore, at the design stage, maintenance requirements should always be considered. Luminaires should be easily accessible for maintenance and lamp replacement. Bulk lamp replacement should be planned so that they are replaced before their useful life expires and their light output deteriorates to an unacceptable level. Lamp and luminaire cleaning should be planned similarly. To minimise disruption, planned cleaning and lamp replacement can take place during holidays.

3.4.4.6 Tips for energy conservation in lighting systems

To effectively save energy in lighting systems, consider the following strategies, organised to emphasise practical implementation and efficiency:

1. Maximise Daylight: Use natural light during the day to reduce the need for electric lighting. This not only saves energy, but also improves the ambience.
2. Choose efficient lighting sources: Choose the most efficient lamps available, as they produce more light per watt consumed, resulting in significant energy savings.
3. Implement effective luminaires. Use luminaires that provide enough light for specific tasks while increasing overall lighting efficiency. This ensures that lighting is both efficient and economical.
4. Enhance with reflective surfaces. To increase light reflectance in the space, paint the ceilings, walls, and floors, as well as the furniture and equipment, in lighter colours. This can greatly reduce the amount of artificial light required.

5. Integrate heat-transfer luminaires: Use luminaires that can effectively transfer heat, using the heat generated by lights in colder weather and dissipating it in warmer weather.
6. Turn off lights when they are not in use. Develop the habit of turning off lights when they are not needed. This simple action can result in significant energy savings over time.
7. Plan for regular maintenance. Maintain your lighting systems on a regular basis, including keeping lights clean and ensuring that lighting levels are adequate to avoid over-illumination.
8. Design systems for specific activities. Customise your lighting systems to reflect the anticipated activities in each building area. This focused approach allows for more precise lighting that meets specific needs while conserving energy.

3.4.4.7 Checklist for lighting systems for energy conservation

This checklist provides a detailed framework for assessing and improving the energy efficiency of lighting systems in a building.

1. Lighting utilisation

Item to check:

Examine the building to see if lights are left on in unoccupied areas and whether there are enough switches for individual light control.

Corrective action:

Install signage in rooms to remind occupants to turn off lights when they leave.

Organise workspaces to cut down on unnecessary lighting.

Rewire switches to allow for more localised lighting control, rather than controlling large areas with a single switch.

Use timers or photoelectric controls to automatically turn off lights when not in use.

2. Excessive lighting Levels

Item to check:

- Use a light metre to determine whether the lighting levels in each area are adequate and not excessively bright.

Corrective action:

- Turn off or remove excess lighting fixtures to achieve proper illumination levels.

- During the day, turn off lights near windows or open spaces.
- Replace existing lamps with lower-wattage alternatives.
- Encourage the use of task-specific, portable lighting sources such as desk lamps.

3. Outside lighting

Item to check:

- Examine external lighting to ensure it is only used for necessary safety and security purposes.

Corrective action:

- Remove all unnecessary outdoor lighting.
- Replace existing outdoor lighting with energy-efficient, low-wattage alternatives that still provide adequate security.
- Install timers or photoelectric sensors to control lights based on daylight or occupancy, ensuring they only turn on when necessary.
- Upgrade incandescent lights to more energy-efficient LEDs.

4. Cleanliness of lighting fixtures

Item to check:

- Inspect lighting fixtures and lamps for dirt or discoloured shields.

Corrective action:

- To ensure maximum lighting efficiency, clean lamps, and fixtures on a regular basis.
- Replace any discoloured or worn shields.
- Replace old or inefficient luminaires with more modern, easy-to-maintain options.

5. Replacement of lamps

Item to check:

- Verify the wattage of lamps and ballasts, especially those that have burned out, to ensure they meet the minimum requirements for their application.

Corrective action:

- Replace old fluorescent lamps with newer, more energy-efficient, and low-wattage units.
- Upgrade to more efficient ballasts that use less power.

References

[1] IEA (2023), Energy Efficiency 2023, IEA, Paris https://www.iea.org/reports/energy-efficiency-2023,Licence:CCBY4.0. (Accessed: 08 May 2024)

[2] IEA (2023), World Energy Outlook 2023, IEA, Paris https://www.iea.org/reports/world-energy-outlook-2023,Licence:CCBY4.0(report);CCBYNCSA4.0(AnnexA). (Accessed: 08 May 2024)

[3] Barney L. Cape hart, Wayne C. Turner (2016), Guide to Energy Management, Fairmont Press, ISBN: 1420084895

[4] Clive Beggs (2002), Energy: Management, Supply and Conservation, Butterworth Heinemann, ISBN: 0750650966

[5] Wayne C. Turner & Steve Doty (2012), Energy Management Handbook, 7th Ed., Fairmont Press, ISBN: 142008870X

[6] Paul W. O'Callaghan (1993), Energy Management, McGraw-Hill Professional, ISBN: [0077076788].

[7] Mehta, V.K.& Rohit Mehta (2006) Principles of power system. New Delhi: S. Chand & Company.

[8] Solar Tubes: A Smart Way to Brighten Dark Rooms. | Electrical Engineering in Kenya (2016) www.eeekenya.com. Available at: https://www.eeekenya.com/solar-tubes-a-smart-way-to-brighten-dark-rooms/ (Accessed: 16 May 2024).

4

Energy Audits and Surveys

Chapter Preview

This chapter starts by defining energy audit and survey and discussing in detail the main types of energy audits (preliminary, targeted, and comprehensive audits). It also includes a typical audit report template and also identifies energy saving opportunities.

4.1 Introduction

Before implementing any energy-saving measures within an organisation, comprehensive energy data must be collected through an auditing process. Before any energy 'problems' can be addressed, it is necessary to assess the current state of a facility's or organisation's energy consumption and, as a result, diagnose any existing issues. This necessitates conducting an energy audit and analysing the data gathered.

An energy audit is a feasibility study to determine the cost of energy inputs and flows within a facility or organisation over a specific time period. An energy audit aims to identify cost-effective energy measures that reduce operational costs.

Energy audits typically involve collecting data from invoices and metres, surveying plant, equipment, and buildings, and gathering information from managers and other stakeholders. The auditing process should identify opportunities to improve an organisation's operational efficiency and reduce maintenance costs. Furthermore, the procedure should aid in resolving any occupant-comfort issues that may exist.

An energy audit should be viewed as the 'foundation' upon which any future energy management programme is built.

Energy management programmes require continuous monitoring and targeting of energy consumption. To set targets and monitor effectively, organisations must first establish baseline energy consumption, operational patterns, building and equipment condition, and energy-management opportunities to reduce costs.

This information can only be obtained by conducting a comprehensive energy audit on an organisation's facilities.

An energy audit should identify issues that require immediate direct action as well as more in-depth investigation. It should also generate data that can be used to justify future capital investment and raise overall organisational awareness of energy conservation issues. Energy audits provide organisations with both direct and indirect financial benefits. Direct benefits include energy cost savings through consumption reduction or fuel type changes. The indirect benefits are less obvious; reduced maintenance costs will result from improved plant utilisation and fewer operating hours. Furthermore, improved plant utilisation may eliminate excess plant capacity, resulting in lower capital expenditure.

The auditing process should identify energy-management opportunities that, if implemented, will result in financial benefits to an organisation. The magnitude of these financial benefits is not necessarily determined by the level of capital investment. Implementing 'no cost' or low-cost measures, such as changing energy tariffs, rescheduling production activities, and adjusting plant controls, can result in significant cost savings.

Encouraging good housekeeping and investing in small capital items like thermostats and time switches can help reduce energy waste.

Although much can be achieved through low-cost measures, it is sometimes necessary to undertake more capital-intensive measures, such as replacing worn-out machinery or installing a building management system (BMS).

4.2 Types of Energy Audits

They are broadly classified into:

1. Preliminary
2. Targeted
3. Comprehensive audits

Each type is distinguished by the level of detail involved and the depth of the analysis undertaken. It is important to select the appropriate audit type for the facility concerned.

Comprehensive audits

These include detailed energy surveys of plant, equipment, and building fabric, which are time-consuming and costly. They, therefore, should not be taken lightly. It is often preferable to focus detailed surveys on problem areas identified by a preliminary energy audit; otherwise, significant time and money can be wasted. A preliminary audit and the methodical application of a variety of simple analysis techniques can frequently identify major energy problems without the need for costly and detailed energy surveys. These audits provide detailed data on energy inputs and flows within a facility or organisation. They should create detailed energy project implementation plans. Such audits include detailed energy surveys and may make use of complex energy simulation computer software.

Preliminary energy audits

These aim to determine the quantity and cost of each type of energy used in a facility or organisation. They are relatively quick and are intended to assess a project's potential; more detailed energy audits and surveys can always be conducted later if necessary. Preliminary audits are primarily concerned with gathering data from energy invoices and metre readings for a specific time period, which is often the most recent fiscal year. Because such audits are primarily concerned with collecting data from bills and invoices, it can be useful to think of preliminary audits as financial energy audits.

Targeted energy audits

These frequently result from preliminary audits. They provide data and detailed analyses for specific targeted projects. For example, an organisation may target its lighting installation or boilers in order to upgrade these pieces of equipment. Targeted audits involve detailed surveys of target subjects, as well as analysis of energy flows and associated costs. They should make recommendations on what action to take.

Energy audits are typically performed by specialist consultants or energy service companies, rather than in-house staff. Energy service companies make money by offering performance contracts that guarantee the organisation's energy cost savings in exchange for negotiated fees. The main interest of energy service companies is not in the audit itself, but in implementing and managing the plant in accordance with their recommendations.

Some companies may even arrange financing for such projects. When using an energy service company, keep in mind that they have a vested interest in the outcome of any energy audit and may not be completely objective. In contrast, energy consultants are independent and should provide objective advice.

4.2.1 Audit costs

Energy audits can be costly undertakings. The cost of an audit increases as it becomes more complex. It is thus critical to choose the appropriate level of audit for any given application.

Audit costs vary depending on the facility's complexity. Complex facilities, such as hospitals or universities, cost more to audit than, say, schools. The age of the facility may also influence the cost. For example, if a mechanical system is complex and the 'as built' drawings are out of date or unavailable, energy auditors may need to create schematic drawings.

This can be time-consuming and significantly increases audit costs. Given the cost, it is critical that organisations assist their auditors by preparing for the audit ahead of time and providing as much relevant information as possible.

Collect all energy bills, fuel invoices, metre readings, and operational notes, as well as any relevant system or building drawings. When conducting an energy audit, organisations should notify their management team and schedule meetings with key managers and stakeholders.

4.3 Why Is Energy Wasted?

1. Poorly designed structures and installations. Buildings may be poorly insulated, resulting in high space-heating costs, or mechanical ventilation ducts may be undersized, causing excessive fan power consumption.
2. Inadequate control systems. Heating systems may be installed without optimal start control.
3. Poor control settings. Time clock controllers may be set incorrectly, causing buildings to be heated even when they are not in use.
4. Inefficient plant operation is frequently caused by the use of outdated technology, which is exacerbated by poor maintenance practices.
5. Poor operational and working procedures. Lights are often left on in buildings when they should be turned off.

Although the reasons for energy waste are multifactorial, some of the main reasons are as follows:

1. Building designers do not pay energy bills. Designers often choose low-capital-cost solutions, which can lead to increased operating costs. This situation is exacerbated by the fact that the budgets for constructing and operating a facility are typically completely separate.
2. People take energy consumption for granted. Most building occupants and users do not pay their energy bills. Individuals prioritise personal comfort over environmental impact.
3. The majority of organisations lack an energy-efficiency culture.
4. In many countries, energy costs are lower than labour costs.

The list above shows that poor strategic and operational management, as well as a lack of energy-saving culture among staff, lead to significant energy waste.

Energy can often be saved at no cost by improving maintenance procedures and implementing good work practices. This is commonly referred to as 'good housekeeping' and consists of simple measures such as encouraging employees to turn off lights when they are not needed. It is also important to establish good maintenance procedures. It is estimated that by implementing good energy-management practices, organisations can reduce their energy bills by around 20%. It is therefore critical that the human and management aspects of energy consumption be investigated in any energy audit. Any organisation will struggle to achieve long-term energy savings unless it has a supportive management culture.

4.4 Preliminary Energy Audits

Preliminary audits aim to quantify and cost each type of energy input to a facility or organisation over time. Identify where energy is being used within the organisation. The primary procedures involved in such an audit are:

- Collecting data
- Analysing data
- Presenting data
- Setting priorities and making recommendations

At the outset of any audit process, it is critical to collect preliminary data about the geographic location of the facility in question, as well as any relevant distinguishing characteristics such as altitude and orientation. Local weather and degree-day data for the audit period should also be collected. These data will serve as a benchmark against which the facility's energy consumption can be measured. With manufacturing facilities, it will also be necessary to collect data on production output during the audit period, as this has a large impact on energy consumption.

The energy invoice is arguably the most important source of energy data. It is therefore critical that the audit team has all of the relevant energy invoices for the selected audit period. By compiling invoice data, it is possible to create a clear picture of a facility's energy consumption pattern and the associated costs for the various energy inputs. Furthermore, the total amount spent on energy can be calculated from the invoices, indicating the upper limit that can eventually be saved through energy-management measures.

When collecting data from fuel and energy invoices, make sure to collect copies of all utility invoices for the audit period, not just those for which payments were made during that time. It is also critical to gather all invoices or delivery notes pertaining to oil, solid fuel, or liquid petroleum gas for the audit duration. Due to the time lag between delivery and consumption, deliveries made prior to the start of the audit period may also be included. Furthermore, all metering and supply points must be identified from invoices in order to account for all energy inputs.

Any estimated metre readings should be identified because they can produce misleading data. To address the issues with estimated readings, additional invoices should be collected for the same months as the estimated invoice but for years prior to the audit period. These 'real' data can then be compared to the estimated data to determine realistic data for the audit period.

If invoice data is insufficient or unavailable, contact utility companies or fuel suppliers for assistance.

Preliminary analysis of energy invoices can often reveal anomalies that require further investigation. If a relatively small building on a site consumes as much gas as one of its much larger neighbours, something appears to be wrong.

Further investigation can then be conducted, which may reveal that the high gas consumption is caused by the heating plant in the small building running at night when the building is empty.

Given that gathering information from invoices is critical to the auditing process, energy invoices must be understood.

4.4.1 Site records

Larger, more complex facilities, particularly those with an energy manager, frequently keep site energy records. These can be a valuable source of information for an audit team, and they can be used to validate data collected from energy invoices. Sub-meter reading records provide detailed information on energy flows, making them especially useful.

4.4.2 Data analysis

For a preliminary energy audit, focus on techniques that determine:

- the amount and type of energy consumed;
- the facility's performance in relation to other similar facilities, as well as the
- building's characteristic performance.

When all of the energy data has been collected and analysed, it must be compared to various 'yardsticks' for similar facilities.

4.5 Comprehensive Energy Audits

Comprehensive energy audits use essentially the same analysis techniques as preliminary audits, but with a much higher level of detail. Comprehensive audits require detailed energy surveys and sub-metering to accurately determine component energy flows.

Conducting comprehensive energy assessments

Overview

A comprehensive energy assessment is a systematic examination of energy consumption, efficiency, and potential savings within a facility or organisation. These assessments are critical for identifying areas where energy can be optimised, costs reduced, and environmental impact minimised.

Key steps in the process

1. **Initial data collection**:
    - Collect relevant data about the facility, such as its size, layout, and energy-consuming systems (HVAC, lighting, appliances, etc.).

- Obtain historical energy bills to create a baseline for comparison.
- Identify the key stakeholders who will participate in the assessment process.

2. **On-site inspection**:
   - Physically inspect the facility to learn about its energy usage patterns.
   - Document all equipment, insulation, windows, doors, and other components.
   - Collect real-time data with tools like energy metres, thermal imaging cameras, and data loggers.
3. **Energy use analysis**:
   - Calculate the energy consumption for various systems (electricity, gas, water).
   - Compare actual usage to industry benchmarks or standards.
   - Determine which areas use the most energy and where there may be inefficiencies.
4. **Identify opportunities for improvement**:
   - Assess lighting systems, HVAC efficiency, insulation, and appliance usage.
   - Consider the renewable energy options (solar panels, wind turbines).
   - Evaluate behavioural practices (employee habits, operational schedules).
5. **Financial analysis**:
   - Determine the cost of implementing energy-saving measures.
   - Calculate the potential energy savings over time.
   - Determine the payback period and return on investment (ROI).
6. **Stakeholder engagement**:
   - Involve facility managers, employees, and other key stakeholders.
   - Discuss the results, proposed changes, and potential benefits.
   - Obtain buy-in and support for energy-efficiency initiatives.
7. **Recommendations and reporting**:
   - Prepare a detailed report summarising the findings and proposed actions.

- Prioritise recommendations according to feasibility, impact, and cost-effectiveness.
- Include clear communication about the advantages of energy efficiency.

Tools and techniques

1. **Energy metres**:
   - Measure electricity, gas, or water consumption at specific points.
   - Provide real-time data for analysis.
2. **Thermal imaging**:
   - Identify heat loss or insulation gaps.
   - Pinpoint areas of inefficiency (e.g. poorly insulated walls or windows).
3. **Data loggers**:
   - Track energy consumption patterns over time.
   - Effective for identifying trends and anomalies.

Stakeholder involvement and communication

- Engage facility managers, maintenance personnel, and occupants.
- Communicate the findings clearly, emphasising the financial and environmental benefits.
- Encourage behavioural change and ongoing monitoring.

Remember that performing a comprehensive energy assessment is not a one-time event. Regular reviews and adjustments are critical for ensuring energy efficiency and sustainability.

4.5.1 Portable and temporary sub-metering

Installing additional sub-metering during an audit can provide valuable insights into a facility's energy flow. For example, by placing sub-meters on the energy input side and heat metres on the output side, the efficiency of individual plant items can be determined. Sub-metering should also reveal any discrepancies between the consumption recorded by main metres and the total recorded by sub-meters. If additional sub-meters are deemed necessary, the additional cost is usually justified for plant items or areas with high loads, especially when there is little information on current energy consumption.

The process of assessing a facility for additional sub-metering may also reveal flaws in the facility's current metering provision. For example, a single electricity metre may serve multiple buildings. In such a case, it may be worthwhile to consider installing permanent sub-meters to support any future energy management programme.

Permanent metering should be considered if it is less expensive than hiring, installing, and removing temporary metering equipment.

Installing metres, whether permanent or temporary, requires shutting off energy supplies, which is often unsatisfactory. Instead, consider the use of portable, non-invasive metering.

4.5.2 Estimating energy use

In many cases, installing comprehensive sub-metering is either impractical or prohibitively expensive, so energy consumption of various plant and equipment items must be estimated. Accurate estimation of equipment energy consumption can be a difficult task that requires skill and judgement. Nonetheless, eqn (4.1) makes it relatively simple to establish an upper limit for energy consumption.

$$\text{Annual energy consumption (kWh)} = \frac{Q_{\text{out}}}{\eta} \times T_h, \tag{4.1}$$

where Q_{out} is the plant power output (kW), η is the efficiency of the plant, and T_h is the number of operating hours per year.

Eqn (4.1) may provide an upper limit on plant energy consumption, but it does not provide actual operating energy consumption. This issue can be solved by measuring actual plant energy consumption for a short period of time using metres and then multiplying the average measured load by the annual operating time. This can be a relatively simple procedure for electrical equipment because the current can be measured with a portable clamp-on metre.

Space-heating energy consumption can be estimated using the heat loss and degree day methods *(beyond scope of this book)*.

4.6 Energy Surveys

Energy surveys are an essential component of the auditing process. They help auditors understand facility energy flows and identify waste.

Surveys can be either comprehensive, looking at all aspects of a facility's energy consumption, or targeted, covering only specific issues. The primary goals of an energy survey should be to:

- determine a facility's energy performance or, in the case of a targeted survey, the specific plant and equipment items;
- identify and quantify key energy flows;
- determine and quantify achievable energy cost savings;
- create cost-effective recommendations to reduce energy costs;
- make recommendations for the facility's future energy management.

Except for specifically targeted surveys, energy surveys should address all aspects of a facility's or organisation's energy consumption. This will include detailed surveys of:

- the management and operational characteristics of a facility or organisation;
- the energy supply for an organisation's various facilities;
- energy consumption within a facility;
- the plant and equipment inside a facility;
- the structure of the organisation's buildings.

4.6.1 Management and operating characteristics

An organisation's management culture has a significant impact on energy consumption. It is therefore critical to determine the management structure and practices relating to energy procurement and consumption.

It is especially important to clearly identify cost centres where managers in charge of operating costs can be held individually accountable for energy consumption. Maintenance practices can also have a direct impact on energy consumption, so it is critical to determine the frequency and quality of maintenance procedures, as well as to identify new maintenance measures that can improve the energy performance of plants and equipment.

During the auditing stage, it is critical to examine an organisation's or facility's operating procedures. Detailed information should be collected on factors such as:

- The use of a specific space or building.
- Mechanical and electrical services in the building.
- The number and types of occupants. Any unique characteristics of the occupants should be carefully considered. For example, windows

are frequently opened in rooms with smokers, resulting in higher space-heating costs.

- The occupancy patterns of a structure or space.
- The environmental conditions of a space or building. This will include air temperature, dry resultant temperature, relative humidity, and light levels.
- The operating procedures for major pieces of plant and equipment.

4.6.2 Energy supply

Identifying the tariffs and supply contracts for an organisation's energy purchases is crucial. This allows the energy auditing team to determine whether or not a specific organisation is purchasing energy at a low cost. If an organisation is overpaying for energy, auditors may suggest switching to a different supplier or tariff.

Choosing the right electricity tariff for an organisation's load profile is crucial, as demand charges are typically included.

As a result, the audit should include a survey of a facility's electrical load profile.

For relatively minor loads, it may be sufficient to take metre readings at the start and end of a specified period, with intermediate readings taken during the day, night, and weekend. This can help auditors recommend appropriate tariffs by providing a clear indication of when electrical energy is consumed. Larger electrical loads require an accurate survey of the load profile. This can be accomplished by using a portable metre to measure demand and consumption at 30-minute intervals over a set time period. Any significant load peaks should be identified, and additional investigations should be conducted to determine their cause.

When it comes to electricity supply, determining a facility's power factor is critical. Equipment like fluorescent lamps and electric motors often have a low power factor (i.e., a decoupling of the current and the voltage so that they become out of phase with each other). This leads to higher-than-expected electricity bills. If a facility has poor power factors, it may be worthwhile to install power factor correction equipment.

4.6.3 Plant and equipment

Major plant components, such as boilers and refrigeration chillers, transfer energy from one form to another. This wastes energy. In boilers, excess heat from combustion can be wasted by escaping with flue gases. The more efficient a plant, the less energy is lost. Major pieces of plant and equipment should be surveyed to determine their operational efficiency. It is also necessary to inspect their respective pipe distribution networks, as these can be a significant source of energy waste.

Boilers should be tuned to reduce flue gas heat losses. This involves sampling the CO_2 or O_2 content of the flue gases and adjusting the burner settings to reduce excess O_2 while ensuring complete combustion. It is important to determine if flue gas heat recovery is feasible.

The coefficient of performance measures a refrigeration plant's efficiency (COP). The higher the COP, the more efficient the machine. COP varies with cooling load and ambient air conditions. To determine the average COP, energy input and output must be measured over time. Take note of the refrigeration plant's operating pattern and how it is controlled, as this will tell the auditor a lot about how the facility operates.

Many facilities waste a significant amount of energy from hot water, chilled water, and steam distribution pipework due to insufficient or poor insulation. Pipework systems should be inspected to determine insulation quality and identify leaks.

Plant surveys should account for the fact that mechanical equipment has a limited working life and that efficiency frequently deteriorates when a plant ages. As a result, one of the key findings of such a survey should be a recommendation for the planned replacement of older plants. In many cases, replacing the old plant is far more cost-effective than renovating it.

4.6.4 Building fabric

It is critical to consider the age, size, shape, and orientation of the buildings within a facility, as these all influence energy consumption. Specifically, areas with the greatest heat loss should be identified.

When inspecting buildings, keep in mind that excessive ventilation can cause significant heat loss.

Particular attention should be paid to any poorly fitting window and door frames or to any space where windows and exterior doors remain open for any length of time.

4.7 Recommendations

The energy auditing process should allow recommendations to be made, resulting in cost savings. Although the precise nature of these recommendations will depend on the particular application in question, they can generally be classified as follows:

1. Energy costs can be reduced by negotiating tariffs for electricity and gas supply. Not all tariffs and supply contracts are suitable for all organisations, and some are superior to others. Changing tariffs or negotiating a better supply contract could help reduce energy costs significantly.
2. Effective maintenance and work practices: 'Good housekeeping' can often result in energy savings at no cost (i.e., improving maintenance procedures and implementing good work practices).
3. Retrofitting and tuning systems: As systems age, components wear out or become damaged, resulting in energy waste. Furthermore, the controls associated with these systems are frequently ineffective or incorrectly configured, resulting in inefficient performance. Significant energy savings can be achieved with a small capital investment to retrofit and re-tune inefficient installations.
4. Capital investment: In many cases, the poor condition of plants and infrastructure renders refurbishment unnecessary. Under these circumstances, a significant capital investment is required to replace the existing plant. In this case, reassessing the situation is often worthwhile to determine whether a different installation would be more appropriate. A combined heat and power (CHP) plant could replace an existing boiler system, reducing the need to purchase electricity. Such measures typically involve significant capital investment and necessitate careful financial analysis.

4.8 The Audit Report

The audit should identify potential energy-management opportunities. Because taking advantage of these opportunities frequently requires capital investment, the greatest effort should be directed toward identifying the measures that will result in the greatest cost savings. Energy-management opportunities that result in smaller savings should be given low priority.

The final audit report should contain:

1. Provide a facility description, including layout drawings, construction details, operating hours, equipment lists, and relevant materials and product flows.
2. An overview of utility tariffs and contracts.

3. A presentation of all the energy data collected, along with any relevant analyses.
4. A detailed statement of potential energy-management opportunities, along with supporting cost/benefit analysis calculations.
5. An energy management action plan for the facility's future operations. This may include a schedule for implementing the recommended energy-management opportunities, as well as a programme for ongoing facility energy monitoring and targeting.

Although the audit report should include detailed technical information, it is important to remember that its primary purpose is to communicate the audit is key findings to an organisation's senior management, many of whom may have little knowledge of energy issues. It is therefore advisable to include a short executive summary, which provides a brief synopsis of the report and highlights its main findings and recommendations.

4.8.1 Detailed energy audit report template

Table of contents

i. Acknowledgement

ii. Executive summary

Energy audit options at a glance and recommendations

1.0 Introduction about the plant

1.1 General plant details and descriptions
1.2 Energy audit team
1.3 Component of production cost (raw materials, energy, chemicals, manpower, overhead, others)
1.4 Major energy use and areas

2.0 Production process description

2.1 Brief description of manufacturing process
2.2 Process flow diagram and major unit operations
2.3 Major raw material inputs, quality and costs

3.0 Energy and utility system description

3.1 List of utilities
3.2 Brief description of each utility

3.2.1 Electricity

3.2.2 Steam
3.2.3 Water
3.2.4 Compressed air
3.2.5 Chilled water
3.2.6 Cooling water

4.0 Detailed process flow diagram and energy and material balance

4.1 Chart showing flow rate, temperature, pressures of all input output streams
4.2 Water balance for entire industry

5.0 Energy efficiency in utility and process systems

5.1 Specific energy consumption
5.2 Boiler efficiency assessment
5.3 Thermic fluid heater performance assessment
5.4 Furnace efficiency analysis
5.5 Cooling water system performance assessment
5.6 Diesel generator set performance assessment
5.7 Refrigeration system performance
5.8 Compressed air system performance
5.9 Electric motor load analysis
5.10 Lighting system

6.0 Energy conservation options and recommendations

6.1 List of options in terms of no cost/low cost, medium cost and high investment cost, annual energy and cost savings, and payback
6.2 Implementation plan for energy saving measures/projects

Annexure

A1. List of energy audit worksheets
A2. List of instruments
A3. List of vendors and other technical details

4.9 Energy Audit Checklist for Building Systems

This checklist is designed to guide energy auditors through a comprehensive evaluation of various building systems to identify potential energy-saving opportunities and improve efficiency.

1. **Lighting systems audit**
   - Lighting inventory: Count the number and type of lighting fixtures in each area.
   - Light measurement: Measure and document foot-candle levels in various areas to determine compliance with recommended levels.
   - Switching mechanisms: Examine the methods used to control lighting, such as manual switches and automatic sensors.
   - Usage patterns: Calculate the average lighting usage hours in each area.
   - Compliance check: Examine and compare current lighting practices to the most recent Illuminating Engineering Society (IES) guidelines.
2. **Building utilisation audit**
   - Space utilisation: Determine how each building area is used, with a focus on primary functions.
   - Operational hours: Record the hours of operation for each area.
   - HVAC needs: Determine HVAC requirements based on area usage.
   - Occupancy patterns: Estimate and record the number of occupants throughout the day and how they vary over the course of 24 hrs.
3. **Building Envelope survey**
   - Structural inspection: Examine the walls, roofs, floors, ceilings, and entryways for integrity and potential energy loss.
   - Insulation assessment: Check the adequacy and condition of insulation in all key areas.
   - Examine and document potential air infiltration points around doors, windows, and other openings.
4. **Electrical systems audit**
   - Utility bill analysis: Examine electricity bills for demand charges to determine the implications of peak usage.
   - Review the utilisation schedule for major electrical equipment.
   - Transformer check: Examine the transformer ratings and actual loading to determine efficiency.
   - Power factor analysis: Determine the power factor of the electrical system and identify areas for improvement.
   - Motor evaluation: Evaluate the loading conditions for all significant electric motors.

5. **Steam and hot water systems audit**
    - System condition review: Inspect the steam and hot-water systems.
    - Boiler efficiency testing: Run tests to determine boiler efficiency.
    - Utilisation patterns: Record the daily and seasonal use of steam and hot water systems.
    - Insulation check: Inspect the insulation in pipes and distribution systems to ensure it is adequate and in good condition.
6. **HVAC systems audit**
    - Temperature settings: Determine the optimal temperatures for energy-efficient operation during different seasons.
    - Air handling systems: Check the exhaust fans, as well as the supply and return air systems, for proper operation.
    - Outdoor air analysis: Track the amount of outdoor air used and how it affects HVAC efficiency.
    - Efficiency tests: Conduct tests to ensure that the HVAC systems are operating at their peak efficiency.
    - Standards compliance: Review and ensure adherence to ASHRAE Standard 90-75.
7. **Special-purpose systems audit**
    - Check and record the temperature settings of domestic hot water systems.
    - Process equipment usage: Examine the utilisation rates and operating schedules of special-purpose process equipment.

Additional notes

- Record keeping: Keep detailed records for each audit area to track changes and progress over time.
- Long-term monitoring: Some systems may require extended monitoring to accurately assess usage patterns and efficiency across multiple seasons or operational cycles.

4.10 Instruments and Metering for Energy Audit

The requirement for an energy audit to identify and quantify where energy is being used necessitates measurements. These measurements require the use of instruments. The basic instruments used in energy audit are as described below.

Key performance parameters for energy audit

Basic electrical parameters in AC and DC systems – voltage (V), current (I), power factor, active power (kW), maximum demand (kVA), reactive power (kVAR), energy consumption (kWh), frequency (Hz), harmonics, etc.

Parameters of importance other than electrical such as temperature and heat flow, radiation, air and gas flow, liquid flow, RPM, air velocity, noise and vibration, dust concentration, TDS, PH, moisture content, relative humidity, flue gas analysis – CO_2, O_2, CO, SO_X, NO_X, combustion efficiency, etc.

Electrical measuring instruments

These are instruments for measuring major electrical parameters such as kVA, kW, PF, Hertz, kVAR, Amps, and Volts. In addition, some of these instruments also measure harmonics (Figure 4.1).

These instruments are applied on-line, i.e., on running motors without any need to stop the motor. Instant measurements can be taken with hand-held metres, while more advanced ones facilitate cumulative readings with print outs at specified intervals.

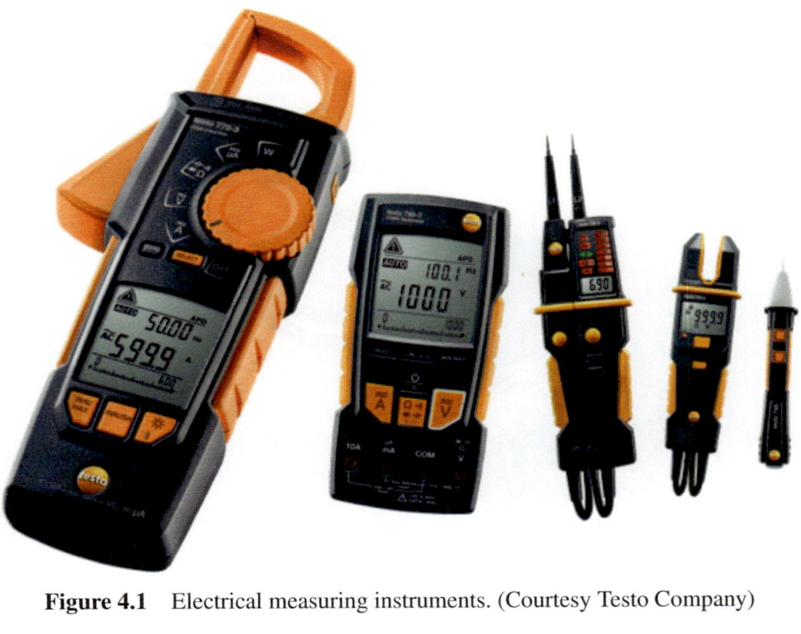

Figure 4.1 Electrical measuring instruments. (Courtesy Testo Company)

Combustion analyser

This instrument has in-built chemical cells which measure various gases such as O_2, CO, NO_X, and SO_X (Figure 4.2).

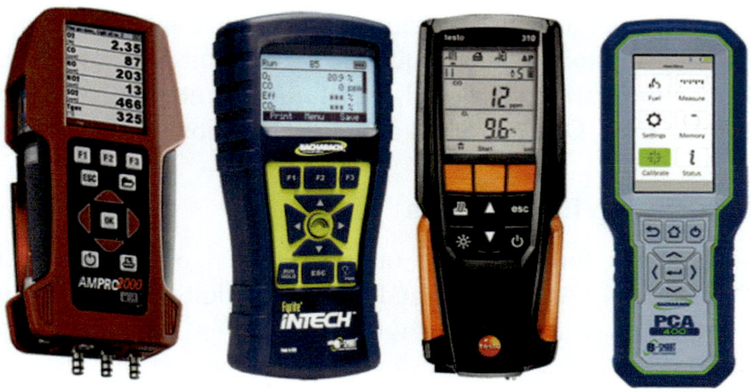

Figure 4.2 Combustion analyser. (Courtesy of InTech, Testo, and Bacharach)

Fuel efficiency monitor

This measures oxygen and temperature of the flue gas. Calorific values of common fuels are fed into the microprocessor which calculates the combustion efficiency (Figure 4.3).

Figure 4.3 Fuel efficiency monitor. (Courtesy of Flowquip Company)

Fyrite

A hand bellow pump draws the flue gas sample into the solution inside the fyrite. A chemical reaction changes the liquid volume revealing the amount of gas. A separate fyrite can be used for O_2 and CO_2 measurement (Figure 4.4).

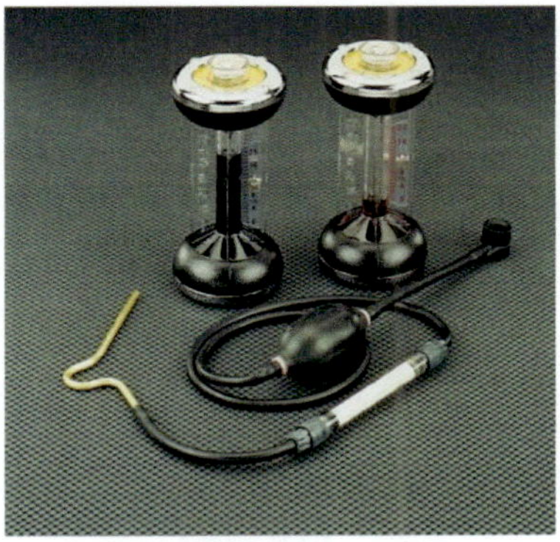

Figure 4.4 Fyrite. (Courtesy of the Bacharach Company)

Contact thermometer

These are thermocouples which measures for example flue gas, hot air, hot water temperatures by insertion of probe into the stream (Figure 4.5).

For surface temperature, a leaf type probe is used with the same instrument.

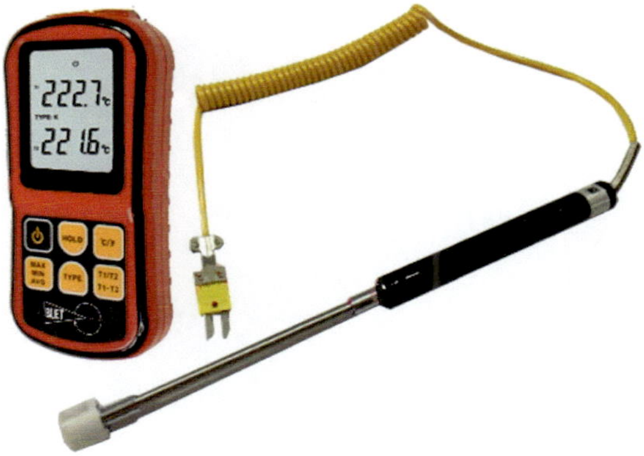

Figure 4.5 Contact thermometer. (Courtesy of BLET Measurement Group)

Infrared thermometer

This is a non-contact type measurement which when directed at a heat source directly gives the temperature read out (Figure 4.6). This instrument is useful for measuring hot spots in furnaces, surface temperatures, etc.

Figure 4.6 Infrared thermometer. (Courtesy of HIOKI Company)

Thermography

Infra-red thermal monitoring and imaging (non-contact type) measures thermal energy radiation from hot/cold surfaces of an object and provides input for assessing health of equipment and predictive maintenance (Figure 4.7).

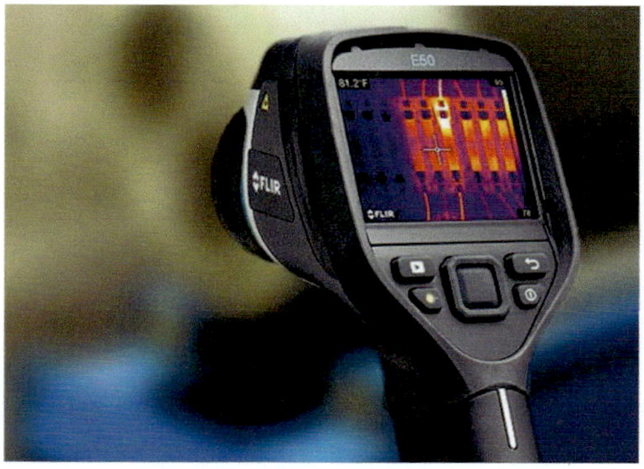

Figure 4.7 Thermography. (Courtesy of Teledyne FLIR LLC Company)

Pitot tube and manometer

Air velocity in ducts can be measured using a pitot tube and inclined manometer for further calculation of flows (Figure 4.8).

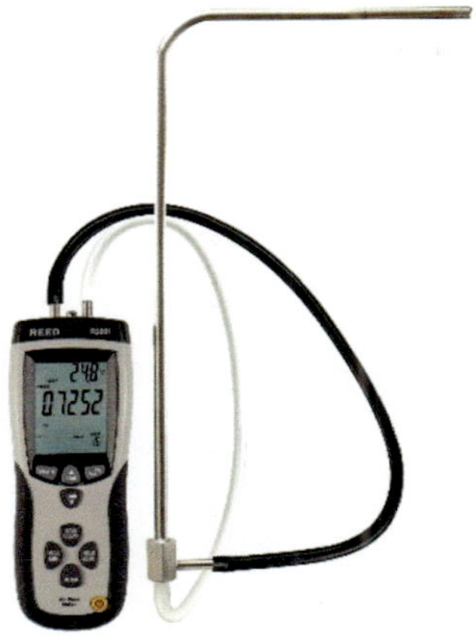

Figure 4.8 Pitot tube and manometer. (Courtesy of Reed Instruments)

Water flow metre

This non-contact flow measuring device uses Doppler effect/ultrasonic principle (Figure 4.9).

There is a transmitter and receiver which are positioned on opposite sides of the pipe. The metre directly gives the flow. Water and other fluid flows can be easily measured with this metre.

Figure 4.9 Water flow metre. (Courtesy of the Walfront Company)

Speed measurements

In any audit exercise speed measurements are critical as they may change with frequency, belt slip, and loading, etc.

A simple tachometer is a contact type instrument which can be used where direct access is possible.

Figure 4.10 Speed measurements. (Courtesy of Uni-Trend Technology Co., Ltd)

More sophisticated and safer ones are non-contact instruments such as stroboscopes (Figure 4.10).

Leak Detectors

Ultrasonic instruments are available which can be used to detect leaks of compressed air and other gases which are normally not possible to detect with human abilities (Figure 4.11).

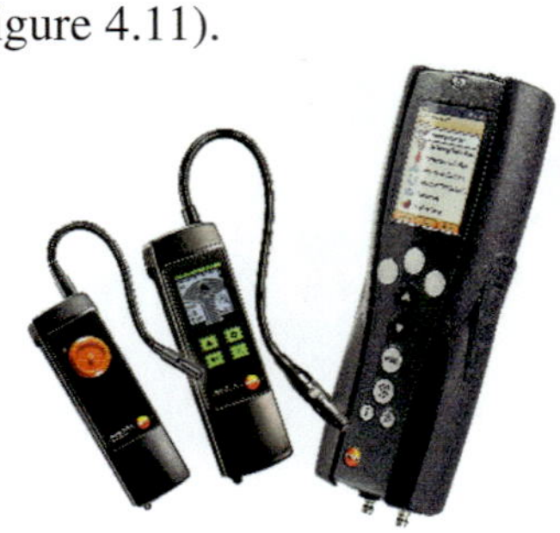

Figure 4.11 Leak detectors. (Courtesy of Testo Company)

Lux metres

Illumination levels are measured with a lux metre. It consists of a photo cell which senses the light output, converts to electrical impulses which are calibrated as lux (Figure 4.12).

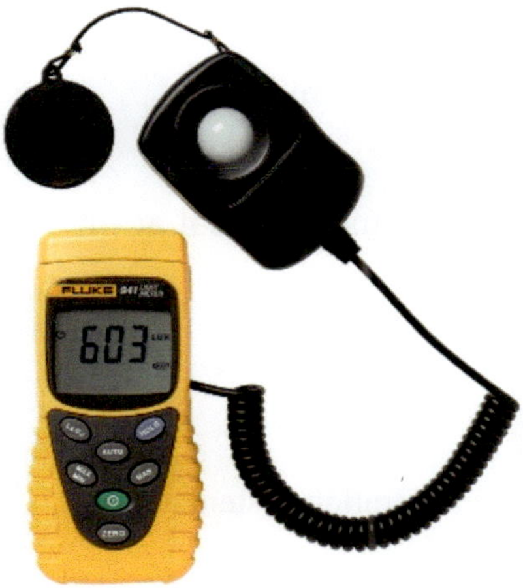

Figure 4.12 Lux metres. (Courtesy of the Fluke Company)

4.11 Additional Notes

4.11.1 ISO standards for energy audit

1. ISO 9001:2015 Quality management systems – Requirements
2. ISO 14001:2015 Environmental management systems – Requirements with guidance for use
3. ISO 45001:2018 Occupational health and safety management systems – Requirements with guidance for use
4. ISO 22000 – Food safety management
5. IATF 16949:2016 – International standard for automotive quality management systems.
6. ISO 50001:2018 – Energy management systems – Requirements with guidance for use

4.11.2 Areas covered under electrical audit

1. Electrical system

2. Electrical distribution system (substation and feeders study)
3. PF improvement study
4. Capacitor performance
5. Transformer optimisation
6. Cable sizing and loss reduction
7. Motor loading survey
8. Lighting system
9. Electrical heating and melting furnaces
10. Electric ovens

4.11.3 Areas covered under mechanical audit

1. Mechanical systems
2. Fans and blowers
3. Exhaust and ventilation system
4. Pumps and pumping system
5. Compressed air system
6. Air conditioning and refrigeration system
7. Cooling tower system

4.11.4 Areas covered under thermal energy audit

1. Thermal energy system
2. Steam generation boilers
3. Steam audit and conversation
4. Steam trap survey
5. Condensate recovery system
6. Insulation survey
7. Energy and material balance for unit operation
8. Heat exchanger
9. Waste heat recovery system

4.11.5 Purpose and importance of energy audits

Energy audits are useful in a variety of settings, including:

1. Industrial settings: Concentrate on lowering operational costs and increasing production efficiency through energy optimisation.
2. Commercial buildings: Aim to reduce energy costs, improve the working environment, and enhance the building's green credentials.

3. Residential environments: Assist homeowners in lowering energy bills and increasing home comfort while reducing environmental impact. These audits are critical in developing a strategy for energy conservation and efficiency improvements.

Energy audits serve as essential tools for identifying opportunities to enhance energy efficiency and reduce energy consumption.

1. **Understanding energy consumption patterns**:
   - Energy audits are systematic examinations of energy usage within a facility or building. Auditors learn about energy consumption by analysing historical data, monitoring equipment, and assessing energy flows.
   - These insights assist in identifying patterns, peak usage times, and areas where energy is wasted.
2. **Pinpointing areas for optimisation**:
   - Energy audits reveal inefficiencies such as obsolete equipment, leaks, and excessive consumption.
   - Auditors evaluate lighting systems, HVAC (heating, ventilation, and air conditioning) units, insulation, appliances, and other energy-consuming devices.
   - Organisations can reduce energy waste by identifying specific areas for optimisation.
3. **Cost savings and environmental impact**:
   - Implementing energy-saving measures based on audit findings result in significant cost savings.
   - Reduced energy consumption means lower utility bills and operational costs.
   - Furthermore, energy efficiency promotes environmental sustainability by lowering greenhouse gas emissions.
4. **Types of energy audits**:
   - Preliminary audits are quick assessments that identify potential areas for improvement.
   - Walk-through audits are visual inspections that evaluate energy systems and identify low-cost improvements.
   - Detailed audits are comprehensive assessments that include data analysis, equipment testing, and detailed recommendations.

    - Investment-grade audits: Thorough evaluations of major projects, which are frequently used to secure financing.

5. **Recommendations and action plans**:
    - After conducting an audit, energy professionals make actionable recommendations.
    - These could include upgrading equipment, changing operational procedures, or investing in energy-saving technologies.
    - Organisations can then develop tailored action plans to implement these changes.

4.11.6 Components of an energy audit

A typical energy audit includes several components:

- Data collection involves gathering utility bills and operational data for analysis.
- Facility inspections include site visits to inspect equipment, the building envelope, lighting, and HVAC systems.
- Energy use measurement: Using tools such as thermographic cameras and blower door tests to evaluate energy efficiency. Experts with specialised skills are frequently required to correctly interpret data and provide actionable insights.

4.11.7 Pinpointing areas for energy optimisation

Energy audits identify critical areas for optimisation:

- Lighting: Assessing current lighting systems and looking for more efficient lighting solutions.
- HVAC systems: Evaluating heating, ventilation, and air conditioning to improve efficiency.
- Insulation: Identifying inadequate insulation in the building envelope that causes energy loss. Improvements in these areas can result in significant energy savings.

4.11.8 Energy audits leading to actionable plans

The findings from an energy audit culminate in an actionable plan:

- Prioritised measures: Recommendations are usually prioritised based on their cost-effectiveness and impact on overall energy efficiency.

- Implementation strategy: The plan includes a strategy for carrying out the recommended measures, often with estimated costs and payback periods.

4.11.9 Identifying energy savings opportunities

After conducting an energy audit, the goal is to identify and prioritise measures that can result in significant energy savings. Here are some common opportunities that auditors frequently identify:

1. **Lighting system improvements**:
    - LED lighting: By replacing traditional incandescent or fluorescent bulbs with energy-efficient LED (light-emitting diode) lights, you can significantly reduce energy consumption. LEDs use less electricity, have a longer lifespan, and generate less heat.
    - Occupancy sensors: Installing occupancy sensors in rooms, hallways, and common areas ensures that lights are only turned on when necessary. These sensors detect movement and automatically turn lights on and off.
    - Daylight harvesting: Using natural daylight effectively by adjusting artificial lighting levels based on available sunlight can result in significant energy savings.
2. **HVAC (Heating, ventilation, and air conditioning) optimisation**:
    - Thermostat settings: By properly setting and programming thermostats, you can avoid unnecessary heating or cooling. Adjusting temperatures during non-business hours or when spaces are vacant can help save energy.
    - Regular maintenance of HVAC systems, such as cleaning filters, checking ducts, and calibrating thermostats, improves efficiency.
    - Zoning: Using zoning systems allows different building areas to be heated or cooled independently, thereby optimising energy consumption.
3. **Equipment upgrades**:
    - Energy-efficient appliances: Replacing older appliances (such as refrigerators, water heaters, and office equipment) with more energy-efficient models can result in significant savings.
    - Variable speed drives (VSDs): Installing VSDs in motors and pumps allows them to operate at different speeds, adjusting energy output to demand. This reduces unnecessary energy consumption.

- Insulation and sealing: Proper insulation and sealing of doors, windows, and ducts reduces energy losses, particularly in heating and cooling systems.

4. **Cost-benefit analysis**:
   - Prioritising energy-saving measures necessitates assessing their cost-effectiveness. A thorough cost-benefit analysis helps determine which strategies provide the best return on investment.
   - Consider the initial costs, anticipated energy savings, maintenance expenses, and payback period.

   Conducting a cost-benefit analysis is essential:
   - Evaluating financial impact: Weighing the initial costs against long-term savings to determine the viability of each energy-saving measure.
   - ROI and payback periods: Calculating return on investment and payback periods to prioritise measures with quick returns.

Additional areas of energy savings

Other key areas for energy savings include:

- Insulation upgrades: Increasing wall, roof, and window insulation to reduce heating and cooling requirements.
- Water heating systems: Converting to more efficient water heating systems.
- Renewable energy: Considering incorporating renewable energy sources such as solar or wind power.
- Behavioural changes: Staff training and promotion of energy-efficient practices.

Prioritising energy-saving measures

Prioritisation is based on multiple factors:

- Low-cost, high-impact measures: Initiatives such as behavioural changes can result in immediate savings with minimal investment.
- Long-term strategies: While more expensive, larger projects, such as HVAC upgrades, can provide significant long-term benefits.

References

[1] IEA (2023), Energy Efficiency 2023, IEA, Paris https://www.iea.org/reports/energy-efficiency-2023, Licence: CC BY 4.0

[2] IEA (2023), World Energy Outlook 2023, IEA, Paris https://www.iea.org/reports/world-energy-outlook-2023, Licence: CC BY 4.0 (report); CC BY NC SA 4.0 (Annex A)

[3] Barney L. Cape hart, Wayne C. Turner (2016), Guide to Energy Management, Fairmont Press, ISBN: 1420084895

[4] Clive Beggs (2002), Energy: Management, Supply and Conservation, Butterworth Heinemann, ISBN: 0750650966

[5] Reyes, M. delos (2023) Energy Consultant: A Complete Guide to the Career, Earnings, and Business - Sustainability Education Academy. Available at: https://sustainabilityeducationacademy.com/energy-consultant-a-complete-guide-to-the-career-earnings-and-business/ (Accessed: 8 May 2024).

[6] International Collaboration - Energy System IEA. Available at: https://www.iea.org/energy-system/decarbonisation-enablers/international-collaboration. (Accessed: 8 May 2024).

[7] Building Energy Management Open-Source Software (BEMOSS) Energy.gov. Available at: https://www.energy.gov/eere/buildings/articles/building-energy-management-open-source-software-bemoss. (Accessed: 8 May 2024).

[8] Practical ways to apply data analytics and AI for energy management and emission control in the steel and cement industries (2024) Industrial Software. Available at: https://new.abb.com/industrial-software/digital/energy-managment/practical-ways-to-apply-data-analytics-and-ai-for-energy-management-and-emission-control-in-the-steel-and-cement-industries. (Accessed: 8 May 2024).

[9] Implementing Energy Management Systems. Available at: https://betterbuildingssolutioncenter.energy.gov/sites/default/files/Implementing_Energy_Management_Systems.pdf (Accessed: 8 May 2024).

[10] Step 1.1 Learn energy management system basics, Energy.gov. Available at: https://www.energy.gov/eere/amo/articles/step-11-learn-energy-management-system-basics.

[11] EECBG program notice 23-01 effective date: April 25, 2023 subject: guidance for eligibility of activities under the energy efficiency and

conservation block grant program. Available at: https://www.energy.gov/scep/articles/energy-efficiency-and-conservation-block-grant-eligible-activities-and-program.

[12] Eligibility: who can get funding?, commission.europa.eu. Available at: https://commission.europa.eu/funding-tenders/how-apply/eligibility-who-can-get-funding_en.

[13] IRENA (2020) Renewable energy and climate pledges: Five years after the Paris Agreement, /publications/2020/Dec/Renewable-energy-and-climate-pledges. Available at: https://www.irena.org/publications/2020/Dec/Renewable-energy-and-climate-pledges. (Accessed: 08 May 2024)

[14] Home (2024) International Institute for Energy Conservation (IIEC). Available at: https://iiec.org/ (Accessed: 8 May 2024).

[15] Yarbrough, D.W. (2022) 'Thermal Insulation for Energy Conservation in Buildings', Springer eBooks, pp. 457–496. Available at: https://doi.org/10.1007/978-3-030-72579-2_19. (Accessed: 08 May 2024)

[16] renew (2023) Energy Saving Retrofits: Building Envelope and Insulation, Renew Energy Partners. Available at: https://renewep.com/energy-saving-retrofits-building-envelope-and-insulation/ (Accessed: 8 May 2024).

[17] Building Envelope Better Buildings Initiative (2023) Energy.gov. Available at: https://betterbuildingssolutioncenter.energy.gov/building-envelope. (Accessed: 08 May 2024)

[18] Steps in Developing a Research Proposal, open.lib.umn.edu. University of Minnesota Libraries Publishing edition, 2015. This edition adapted from a work originally produced in 2010 by a publisher who has requested that it not receive attribution. Available at: https://open.lib.umn.edu/writingforsuccess/chapter/11-2-steps-in-developing-a-research-proposal/.(Accessed: 08 May 2024)

[19] Renewable Energy and Energy Efficiency Incentives-A Summary of Federal Programs Congressional Research Service. Available at: https://crsreports.congress.gov/product/pdf/R/R40913/29 (Accessed: 08 May 2024).

[20] https://www.energy.gov/scep/energy-efficiency-and-conservation-block-grant-eecbg-program-frequently-asked-questions

[21] Renewable energy explained - incentives - U.S. Energy Information Administration (EIA) (2016) Eia.gov. Available at: https://www.eia.gov/energyexplained/renewable-sources/incentives.php.

[22] El-Samra, S. and Adriazola-Steil, C. (2022) '5 Ways to Cut Oil and Gas Use Through Clean Transportation', www.wri.org [Preprint]. Available at: https://www.wri.org/insights/5-ways-cut-oil-and-gas-use-through-clean-transportation.

[23] 5 ways for the world to reduce emissions from global transport systems (2023) World Economic Forum. Available at: https://www.weforum.org/agenda/2023/03/5-ways-the-world-can-reduce-emissions-from-global-transport-systems/.

[24] INCLUSIVE AND SUSTAINABLE INDUSTRIAL DEVELOPMENT Practical Guide for Implementing an Energy Management System. Available at: https://www.unido.org/sites/default/files/2017-11/IEE_EnMS_Practical_Guide.pdf.

[25] Energy end-use data collection methodologies and the emerging role of digital technologies – Analysis, IEA. Available at: https://www.iea.org/reports/energy-end-use-data-collection-methodologies-and-the-emerging-role-of-digital-technologies.

[26] Evaluate the Economics of Energy Efficiency Projects ENERGY STAR www.energystar.gov. Available at: https://www.energystar.gov/buildings/save-energy-commercial-buildings/economics-efficiency-projects (Accessed: 8 May 2024).

[27] A simple step-by-step guide for ngos on 'how to write proposals' - fundsforngos. Available at: https://www2.fundsforngos.org/featured/a-simple-step-by-step-guide-for-ngos-on-how-to-write-proposals/ (Accessed: 08 May 2024).

5

Renewable and Sustainable Energy Sources

Chapter Preview

In this chapter we review the basic renewable and sustainable energy sources such as solar, wind, hydropower, biomass, and geothermal. We briefly discuss how the different energy systems work, the respective advantages and challenges as well as future trends.

5.1 Introduction

Defining renewable energy

Renewable energy comes from natural processes that are constantly replenished, such as sunlight, wind, water, and geothermal heat. Renewable energy sources, as opposed to non-renewable sources such as coal and oil, provide a sustainable alternative due to their inexhaustibility.

Importance in sustainable energy future

Renewable energy is critical to a sustainable future because it reduces our reliance on fossil fuels and helps to mitigate climate change. Its integration into global energy portfolios is changing the way countries generate power and manage resources.

Misconceptions and challenges

Common misconceptions include concerns about reliability and efficiency. Renewable energy adoption faces several challenges, including initial investment, technology maturity, and infrastructure development.

5.2 Solar Energy

Principles of solar energy generation

Photovoltaic cells convert sunlight into electricity, while solar thermal systems convert sunlight into heat. These technologies use solar power in an efficient and sustainable way.

Applications and technologies

Solar energy has a wide range of applications, from residential rooftop panels to large solar farms. Improved cell efficiency and innovative storage solutions have made solar power more viable and accessible.

Photovoltaic cells

Photovoltaic cells, or solar panels, use the photovoltaic effect to convert sunlight directly into electricity. Illustration of how PV cells produce electricity:

1. Sunlight activation: Sunlight strikes the solar panel cells, and photons are absorbed by a semiconducting material like silicon.
2. Electron excitation: The energy from photons excites electrons in silicon, allowing them to move freely.
3. Electric current generation: The movement of these free electrons towards the front surface of the solar cell causes an imbalance. When electrical conductors are connected to the positive and negative sides, they form an electrical circuit. The flow of electrons through this circuit generates electricity.
4. Power output: The direct current (DC) electricity is then passed through an inverter, which converts it to alternating current (AC) electricity that can be used in homes or fed into the grid.

Solar thermal systems

Solar thermal systems heat a fluid with sunlight, which is then used to generate steam that drives a turbine to generate electricity or to provide direct heat for residential, industrial, or commercial applications. The process includes:

1. Solar thermal panels or mirrors capture solar radiation and focus it to heat a transfer fluid (like water or oil).

2. Heat transfer: The heated fluid is circulated and used to heat water or another working fluid, resulting in steam.
3. Electricity generation: The steam propels a turbine connected to a generator, which generates electricity.
4. Direct use or storage: The heat produced can be used directly to heat buildings or stored in thermal storage systems for later use.

5.3 Wind Energy

Generating wind energy

Wind turbines convert the kinetic energy of the wind into electricity. The design and technology of wind turbines and farms are aimed at increasing efficiency and effectively harnessing wind power.

Advantages and environmental considerations

Wind energy provides a clean, renewable energy source with low environmental impact. However, considerations include the effects on wildlife and land use, which necessitate careful planning and mitigation measures.

How wind power generation works:

1. **Wind capture**: Wind turbines are typically installed in high-wind areas, usually in large groups known as wind farms. When the wind blows, it passes over the turbine's blades and causes them to turn.
2. **Mechanical to electrical conversion**:
    - Blades: Capture wind kinetic energy. They are linked to a rotor, which turns alongside the blades.
    - The gearbox increases the rotor's rotation speed from low to high, making it more suitable for generating electricity.
    - Generator is connected to the gearbox. The gearbox spins the generator, converting mechanical energy into electrical energy via electromagnetic induction.
3. **Power transmission**: The electricity produced is in the form of alternating current (AC). It is transmitted via an underground cable from the turbine to a transformer, which boosts the voltage for long-distance transmission.

4. **Grid connection**: Following voltage transformation, electrical energy is transmitted to the power grid and distributed to consumers.
5. **Control systems**: Modern wind turbines have sophisticated sensors and control systems that optimise the pitch (angle) of the blades and the turbine's orientation (yaw) relative to the wind to maximise efficiency and reduce wear and tear.

Environmental and economic benefits:

- Renewable and clean: Wind energy is a plentiful, renewable resource that reduces reliance on fossil fuels and greenhouse gas emissions.
- Cost-effective: After initial setup costs, wind power has a low operational cost because wind is free. This makes it an affordable energy solution in the long run.
- Land use flexibility: The land surrounding wind turbines can still be used for agricultural purposes.

Challenges:

- Wind is intermittent, which can cause variability in power generation.
- Location dependent: Effective wind farm locations are frequently remote, necessitating extensive infrastructure for electricity transmission.
- Wildlife and noise concerns: Wind turbines can have an impact on local wildlife, particularly birds and bats, as well as generate noise that can be disruptive to nearby residential areas.

Future trends:

- Technological advances: Continuous improvements in turbine technology, materials, and aerodynamics increase wind power efficiency while lowering costs.
- Offshore expansion: Because the winds at sea are stronger and more consistent than that on land, offshore wind farms are rapidly expanding.

5.4 Hydropower

How hydropower works

Hydropower harnesses the energy of moving water to generate electricity. It includes everything from large-scale dam systems to smaller run-of-the-river setups, providing versatile solutions for renewable energy generation.

How hydropower works:

1. Water capture: Hydropower plants typically require a significant water source, such as a river or a man-made reservoir. A dam is frequently built to raise water levels, create reservoirs, or regulate water flow.
2. **Potential and kinetic energy conversion**:
   - **Potential energy**: The water in the reservoir represents potential energy.
   - **Kinetic energy**: As water is released from the reservoir via the dam, it flows downhill through turbines, converting potential energy to kinetic energy.
3. **Electricity generation**:
   - **Turbines**: As water flows through, it turns blades in a turbine, causing it to spin and generate electricity.
   - **Generators**: Generators use electromagnetic induction to convert the turbine's mechanical energy into electrical energy.
4. **Transmission**: Transformers step up the voltage of the electricity generated and transmit it through power lines to homes, businesses, and other areas as needed.

Advantages of hydropower:

- Water is a renewable and sustainable natural resource that is constantly replenished by snow and rain.
- Low operational costs: Once a dam is built, equipment maintenance and operation are relatively inexpensive.
- Reliable energy source: Provides a more consistent source of energy than other renewable energy sources such as solar and wind.
- Storage capability: Reservoirs can store water for later use when electricity demand is high.
- Dams aid in flood management and provide a consistent water supply for agricultural and residential purposes.

Challenges:

- Environmental impact: Dams and reservoirs can have a significant impact on water temperatures, flow regimes, and aquatic ecosystems.
- Displacement: The construction of large reservoirs can displace both human populations and wildlife.
- Dams and other hydroelectric facilities require a significant initial investment.

- Geographical limitations: Suitable hydropower sites are limited to areas with adequate water flow and suitable terrain for reservoirs.

Future trends:

- Pumped storage: This technology raises water to a higher elevation, which is then released to generate electricity when demand is high, serving as an energy storage solution to balance supply and demand.
- Small-scale and micro hydropower: Small-scale hydropower developments enable remote communities to generate power without the use of large dams.

Environmental considerations and future potential

While hydropower is a dependable renewable energy source, it can have environmental consequences for aquatic ecosystems and land use. Ongoing advancements are aimed at minimising these effects while increasing energy output.

5.5 Biomass Energy

Biomass as a renewable source

Biomass energy comes from organic materials such as wood, agricultural residues, and municipal waste. It is a versatile energy source that can be used to generate heat, electricity, and even biofuel for transportation.

How biomass energy is generated:

1. **Source materials**: Biomass energy uses a variety of organic materials:
    - Wood and wood waste include logs, chips, bark, and sawdust from lumber mills.
    - Agricultural crops and waste include corn, sugarcane, and crop residues such as straw and husks.
    - Animal manure and human waste are processed to produce biogas.
    - Municipal solid waste contains organic components that can be converted into energy.
2. **Conversion methods:** Several methods exist for converting biomass into usable energy.

- Direct combustion is the simplest and most common method of burning organic materials to produce heat, which is then used to generate power via steam turbines.
- Anaerobic digestion: In the absence of oxygen, microorganisms break down biodegradable material to produce biogas, primarily methane, which can be burned to generate heat and electricity.
- Fermentation is the process of converting organic materials (particularly sugars) into biofuels such as ethanol, which can then be used to fuel vehicles.
- Pyrolysis is the process of decomposing organic material at high temperatures in the absence of oxygen, producing bio-oil and syngas.

3. **Energy production**:
    - Electricity generation: Biomass can be burned directly in steam boilers or biomass cogeneration plants to generate electricity.
    - Heating and cooking: Biomass is widely used in stoves, particularly in rural areas.
    - Transport fuels: Biofuels like biodiesel and ethanol can be used in vehicles as a clean-burning alternative to fossil fuels.

Advantages of biomass energy:

- Renewable: Derived from plants that can be replenished over a human timescale.
- Carbon neutral: The CO_2 produced when biomass fuel is burned is offset by the CO_2 absorbed during feedstock growth.
- Reduces waste: By reusing materials that would otherwise end up in landfills, overall waste and methane emissions are reduced.

Challenges:

- Crop cultivation is resource intensive, requiring significant land and water resources.
- Competition with food production: Using agricultural crops for energy has the potential to raise food prices.
- Environmental impact: If not managed sustainably, this may result in deforestation. The combustion of biomass emits pollutants such as particulate matter and nitrogen oxides.

Future trends:

- Advanced biofuels: The development of second- and third-generation biofuels which are less competitive with food crops and more energy efficient.
- Integrated bio refineries: Facilities that convert biomass into fuels, power, and chemicals in a single integrated system, increasing efficiency and reducing waste.

Sustainability challenges

The sustainability of biomass energy is determined by factors such as sourcing and energy balance. The challenges include ensuring that it does not harm food supply or biodiversity.

5.6 Geothermal Energy

Harnessing geothermal energy

Geothermal energy harnesses the earth's heat for power generation and heating. Deep-earth drilling and the use of geothermal heat at shallower depths are two different technologies.

How geothermal energy works:

1. Heat source: The earth's core is extremely hot, with temperatures reaching 5000 °C (9032 °F). This heat is constantly transferred to the surface, warming the subsurface rocks and the water that infiltrates them, resulting in a geothermal resource.
2. **Resource extraction**:
    - Hydrothermal reservoirs are the most common type of geothermal resource, as they provide access to naturally occurring pockets of steam or hot water.
    - Enhanced geothermal systems (EGS): Techniques for enhancing or creating geothermal resources by fracturing hot rock and passing water through it to extract heat.
3. **Electricity generation**:
    - Hot water extracted from the ground is used directly to heat buildings, agricultural applications, and industrial processes.

- **Electric power generation**:
    - Dry steam plants use steam directly from a geothermal reservoir to drive a turbine.
    - Flash steam plants: Draw deep, high-pressure hot water into lower-pressure tanks, causing it to vaporise (or 'flash') into steam capable of driving a turbine.
    - Binary cycle power plants: Lower temperature geothermal water is used to heat a secondary fluid with a lower boiling point than water, which is then vaporised and used to drive a turbine.

Applications of geothermal energy:

- Geothermal power stations generate electricity.
- Heating for residential and commercial buildings, including district heating systems.
- Agriculture includes greenhouse heating, fish farming, and crop drying.
- Industrial processes include pasteurisation, drying, and other heat-intensive processes.

Advantages of geothermal energy:

- Sustainability and availability: Offers a consistent, dependable source of energy with minimal fluctuations in output.
- Environmentally friendly: Emits few to no greenhouse gases and has a small footprint when compared to other energy sources.
- Cost-effectiveness: Following an initial capital investment, operational costs are low.

Challenges:

- High initial costs: Exploring and developing geothermal resources incurs significant upfront costs.
- Site-specific: Geothermal plants must be located in areas with access to subsurface heat.
- Environmental concerns: Potential release of harmful gases (such as hydrogen sulphide), land subsidence, and water contamination.

Future trends:

- Drilling and extraction technologies have advanced to allow for more efficient access to deeper and more challenging resources.
- Increased adoption: Geothermal energy is now being used in regions that were previously considered unsuitable due to technological limitations.

Environmental impacts and technological advancements

Geothermal energy is generally environmentally friendly, but it does raise some concerns about land use and potential emissions. Technological advancements continue to improve its efficiency and mitigate potential consequences.

5.7 Integrating Renewable Energy into Existing Systems

Challenges of integration

Integrating renewable energy into existing grids presents challenges such as variability and energy storage. Addressing these issues requires novel solutions, such as advanced battery systems and grid modernisation.

Infrastructure adaptation

Adapting infrastructure to accommodate renewables entails updating grid systems and improving energy storage capabilities to ensure a smooth integration of these renewable energy sources.

References

[1] IEA (2023), Energy Efficiency 2023, IEA, Paris https://www.iea.org/reports/energy-efficiency-2023,Licence:CCBY4.0. (Accessed: 08 May 2024)

[2] Barney L. Cape hart, Wayne C. Turner (2016), Guide to Energy Management, Fairmont Press, ISBN: 1420084895

[3] NCPCAdmin - Case studies - Industrial Energy Efficiency, Industrial Efficiency. Available at: https://www.industrialefficiency.co.za/case-studies/case-studies-iee/ (Accessed: 8 May 2024). renew (2023) Energy Saving Retrofits: Building Envelope and Insulation, Renew Energy Partners. Available at: https://renewep.com/energy-saving-retrofits-building-envelope-and-insulation/. (Accessed: 08 May 2024)

[4] Saving Energy in Industrial Companies: Case Studies of Energy Efficiency Programs in Large U.S. Industrial Corporations and the Role of Ratepayer-Funded Support... Available at: https://www.energy.gov/sites/default/files/2021-07/saving_energy_industrials.pdf (Accessed: 08 May 2024).

[5] Envelope | Better Buildings Initiative (2023) Energy.gov. Available at: https://betterbuildingssolutioncenter.energy.gov/building-envelope. (Accessed: 08 May 2024)

[6] Renewable energy explained - incentives - U.S. Energy Information Administration (EIA) (2016) Eia.gov. Available at: https://www.eia.gov/energyexplained/renewable-sources/incentives.php. (Accessed: 08 May 2024)

[7] Building Envelope | Better Buildings Initiative (2023) Energy.gov. Available at: https://betterbuildingssolutioncenter.energy.gov/building-envelope. (Accessed: 08 May 2024)

6

Energy Conservation in Buildings and Facilities

Chapter Preview

In this we discuss how energy conservation can be integrated in buildings, even at the design stage. We explore the principles of energy-efficient designs, material selection for improved energy performance, integration of renewable energy systems, and energy-efficient HVAC and lighting systems.

6.1 Integrating Energy Management and Conservation at the Design Stage

Introduction

The design phase of a building or system provides an important opportunity to implement energy management and conservation strategies. By incorporating energy-efficient designs, materials, and technologies from the start, developers can significantly reduce a building's environmental impact while also ensuring its operational efficiency for many years. This chapter looks at various strategies and considerations for incorporating energy conservation into the design phase of construction and development projects.

Principles of energy-efficient design

Before delving into specific strategies, it is important to understand the fundamental principles that guide energy-efficient design:

1. Whole-system approach: Viewing the building as an integrated system, with each component working in tandem to maximise energy efficiency.
2. Lifecycle analysis: Examining the energy costs associated with a building or system throughout its entire lifecycle, from construction to operation and eventual decommissioning.

3. Passive solar design: Using the sun's energy for heating and cooling to reduce reliance on external systems.

Key strategies for energy conservation at the design stage

The following strategies can be employed during the design stage to enhance energy conservation:

1. Site and orientation planning: Positioning a building to maximise natural light, heat, and ventilation. This requires careful consideration of the building's orientation, window placement, and landscaping.
2. Building envelope optimisation refers to the design of the building envelope, which includes walls, roofs, and foundations, in order to maintain thermal integrity. This could include using advanced insulation techniques, high-performance windows, and airtight construction methods.
3. Energy modelling is the process of using advanced software to simulate energy usage in various design scenarios in order to identify the most energy-efficient options.
4. Incorporating renewable energy sources: Buildings are designed to incorporate renewable energy systems such as solar panels, wind turbines, or geothermal energy solutions, which can provide clean, low-cost electricity.
5. Choosing the most energy-efficient heating, ventilation, and air conditioning (HVAC) systems and lighting fixtures available. This includes systems such as underfloor heating, LED lighting, and daylight-responsive lighting controls.
6. Water conservation systems: Using water-saving features like rainwater harvesting, greywater systems, and water-efficient fixtures to reduce energy consumption for water heating and management.

Design techniques and tools

Several design techniques and tools can aid architects and engineers in creating energy-efficient buildings:

1. BIM (Building information modelling) software enables the creation of energy models that are integrated into the design process, resulting in a detailed visualisation of energy flows and potential conservation measures.

2. Thermal mass materials: Using materials with high thermal mass in areas with plenty of sunlight can help regulate indoor temperatures by absorbing heat during the day and releasing it at night.
3. Natural ventilation design: Creating building layouts that optimise airflow can significantly reduce the need for mechanical cooling systems.

Case studies

Illustrative examples of successful energy-efficient designs include:

- The Edge in Amsterdam: Known as the world's greenest building, it has an intelligent façade that optimises light and heat efficiency, as well as a sophisticated energy management system.
- The Bullitt Center in Seattle generates more electricity than it consumes each year, making it a net-zero energy building thanks to its solar array and ultra-efficient systems.

Challenges and considerations

While incorporating energy management and conservation into the design process is extremely beneficial, it does present some challenges, such as higher upfront costs and the need for specialised design expertise. Furthermore, achieving energy efficiency goals frequently necessitates close collaboration among architects, engineers, and environmental consultants throughout the design and construction phases.

6.2 Energy-efficient building design and construction

Energy-efficient building design and construction are significant steps towards reducing energy consumption and promoting sustainability. Energy efficiency in building design requires a comprehensive approach that takes into account factors such as orientation, materials, and renewable energy systems. This section looks at how these design choices can result in buildings that are not only less energy-intensive, but also more environmentally sustainable.

6.2.1 Building orientation and design for energy efficiency

The orientation and architectural design of a building play pivotal roles in energy efficiency:

- Maximising natural resources: Proper orientation can increase natural heating, cooling, and lighting, reducing the need for artificial means.
- Architectural elements: The strategic placement of windows, insulation techniques, and shading devices can have a significant impact on a building's thermal performance and energy requirements.
- Why does orientation matter? The direction a building faces has a significant impact on its energy performance. Proper orientation can increase natural light, reduce heat gain, and improve solar energy efficiency.
- **Strategies:**
    - Passive solar design: Aligning the building's longer sides along the east-west axis maximises solar exposure. South-facing windows capture sunlight in the winter and reduce direct sun exposure in the summer.
    - Shading and overhangs: Strategically placing shading devices (such as overhangs and louvres) reduces cooling loads by preventing excessive solar radiation.
    - Site analysis: Use the local climate, prevailing winds, and sun angles to guide design decisions.

6.2.2 Material selection for improved energy performance

Choosing suitable building materials is critical for energy efficiency:

- Thermal performance: Materials such as high-insulation windows and specialty roofing can significantly improve a building's thermal efficiency.
- Sustainable materials: Using sustainable, eco-friendly materials improves energy performance while also lowering the environmental impact of the construction process.
- Why do materials matter? The choice of construction materials has an impact on energy efficiency, indoor air quality, and overall environmental impact.
- **Strategies:**
    - Insulation: Effective insulation reduces heat transfer through walls, roofs, and floors.
    - Thermal mass: Materials with a high thermal mass (e.g. concrete, stone) absorb and release heat slowly, keeping indoor temperatures stable.

- Choose eco-friendly materials such as recycled steel, bamboo, or reclaimed wood.
- Low-E windows: Energy-efficient windows with low-emissivity coatings minimise heat loss and gain.

6.2.3 Integration of renewable energy systems

Incorporating renewable energy systems is key to reducing a building's carbon footprint:

- Solar panels and wind turbines can significantly reduce a building's reliance on non-renewable energy sources.
- Emerging technology: Building-integrated photovoltaic (BIPV) and solar thermal systems are at the forefront of renewable energy integration in building design.
- Design considerations: Cost, aesthetic impact, and system efficiency must all be balanced to ensure effective integration.
- Why switch to renewables? Using clean energy sources reduces our reliance on fossil fuels and lowers greenhouse gas emissions.
- **Strategies:**
    - Solar photovoltaic (PV) panels: Install PV panels on your roof to generate electricity from sunlight.
    - Solar collectors are used to heat water for both domestic and industrial purposes.
    - Wind turbines: In appropriate locations, wind turbines can supplement energy requirements.
    - Geothermal heat pumps: Use the earth's constant temperature for heating and cooling.

6.2.4 Building envelope insulation and optimisation

The building envelope refers to a structure's outer shell, which includes its walls, roofs, windows, and doors. It serves as a protective barrier between the indoor and outdoor environments. When designed and built thoughtfully, the building envelope has a significant impact on energy consumption, comfort, and sustainability.

- Why focus on the envelope? The building envelope (walls, roof, and windows) serves as a barrier between the indoor and outdoor environments.

- **Strategies:**
  - Air sealing: Seal any gaps or cracks to prevent air leaks.
  - High-performance windows: Double-glazed or low-E windows improve insulation.
  - Cool roofs: Reflective roofing materials help to reduce heat absorption.
  - Green roofs: Vegetated roofs increase insulation and absorb rainwater.

Building envelope and insulation

1. **Thermal insulation**:
   - The primary goal of thermal insulation is to reduce heat transfer through the building envelope. It accomplishes this by lowering heat loss during cold weather and increasing heat gain during hot weather.
   - **Materials**: Various insulation materials are available, including:
     - Wool, cotton, and cork are some examples of naturally occurring fibres and particles.
     - Man-made fibres include fibreglass and mineral wool.
     - Reflective systems: Materials that reflect radiant heat.
     - Cellular plastics include materials such as expanded polystyrene (EPS) and extruded polystyrene (XPS).
     - Evacuated systems use vacuum-insulated panels (VIPs) with extremely low thermal conductivity.
     - Aerogels are ultra-lightweight, highly insulating materials.
     - Hybrid Insulation: Using two or more types of insulation to improve performance.
   - Insulation effectiveness is assessed using parameters such as thermal conductivity, R-value, and U-value.
   - Performance limitations: While insulation increases energy efficiency, it is critical to consider its limitations and compatibility with other building materials.
   - Moisture and air infiltration: Proper installation and moisture management are critical for preserving insulation performance.
2. **Impact on energy conservation**:
   - Reduced heating and cooling loads: Well-insulated building envelopes require less energy for heating and cooling.

- Comfort: Adequate insulation maintains constant indoor temperatures and reduces draughts.
- Environmental benefits: By lowering energy consumption, we help to reduce greenhouse gas emissions.
- Financial savings: Energy-efficient buildings reduce costs over time.

3. **Recent advances**:
    - Continuous research and development have resulted in innovations in insulation materials and techniques.
    - Sustainable materials: The trend is to use renewable and environmentally friendly materials.
    - Integration with other elements: Insulation functions in tandem with windows, doors, and roofing systems.

Building envelopes and its role in energy conservation

The building envelope, which includes walls, roofs, windows, and doors, is critical in conserving energy within buildings. It serves as a weatherproof barrier, ensuring interior comfort and reducing energy loss. Understanding how to properly insulate and seal the building envelope is critical to increasing energy efficiency and lowering heating and cooling loads.

Insulation materials and techniques

Insulation is essential in controlling heat transfer:

- Fibreglass, cellulose, and various types of foam are common insulation materials, each suitable for a specific application and building component.
- Advanced insulation: New technologies such as aerogel insulation and vacuum-insulated panels provide superior thermal resistance in thinner profiles.
- Impact on energy efficiency: Proper insulation significantly reduces the need for artificial heating and cooling, resulting in energy savings and lower utility bills.

Sealing and air tightness

Air tightness prevents unwanted air leaks, a major source of energy loss:

- Sealing techniques: Weather-stripping and caulking around windows and doors can significantly improve a building's airtightness.

- Assessment tools: Technologies such as blower door tests aid in identifying and addressing air leaks, resulting in optimal insulation effectiveness.

Energy-efficient windows and doors

Advancements in windows and doors are pivotal in energy conservation:

- High-performance features: Double-glazing, low-emissivity (low-E) coatings, and thermally broken frames all improve energy efficiency.
- Balancing light and heat: These technologies also contribute to maintaining a balance between natural lighting and thermal performance, which is critical in building design.

Roofing and external wall solutions

The external surfaces of a building, especially roofing and walls, significantly impact its energy performance:

- Cool and green roofs: These roofing solutions reflect more sunlight and absorb less heat, lowering cooling requirements.
- External cladding: Cladding systems not only add aesthetic appeal to a building, but they also provide an extra layer of insulation, increasing the overall energy efficiency.

6.2.5 Assessment and retrofitting of existing buildings

Retrofitting existing buildings poses unique challenges, but it is critical for increasing overall energy efficiency. Assessing the current insulation and air tightness can help guide effective retrofitting strategies for increased energy efficiency.

Smart building controls

- Why should we embrace technology? Automated systems improve energy efficiency, increase comfort, and reduce waste.
- **Strategies:**
    - Energy management systems (EMS) monitor and control lighting, HVAC, and appliances.
    - Occupancy sensors: Set lighting and HVAC based on occupancy.

  - Daylight harvesting: Reduce artificial lighting when natural light is sufficient.

6.2.6 Energy-efficient HVAC and lighting systems

High-efficiency HVAC and lighting systems are critical components of energy-efficient buildings.

- HVAC systems: Advanced HVAC systems can significantly reduce energy consumption while still providing indoor comfort.
- Lighting solutions include LED lighting and intelligent lighting controls, which consume significantly less energy than traditional lighting methods.

Lighting and HVAC systems in energy conservation

Lighting and HVAC (heating, ventilation, and air conditioning) systems are among the most energy-intensive components of buildings. Implementing energy-saving strategies in these areas is critical to increasing overall building efficiency and lowering operational costs. This section investigates how improvements in energy-efficient lighting, smart control systems, and optimised HVAC designs can have a significant impact on a building's energy consumption and indoor environmental quality.

Energy-efficient lighting technologies

Advances in lighting technology have opened new avenues for energy conservation.

- LED and OLED lights have higher energy efficiency and longer lifespans than traditional lighting solutions.
  - LED (light-emitting diode) lighting: LEDs are significantly more energy efficient than traditional incandescent bulbs. They use significantly less electricity, have a longer lifespan, and produce less heat.
  - CFL (compact fluorescent lamp) Bulbs: CFLs are another energy-efficient option. They use approximately 75% less energy than incandescent bulbs and last longer.

  - Daylight Harvesting: Using natural daylight effectively reduces the need for artificial lighting during the daylight hours. Sensors can adjust artificial lighting levels in response to available sunlight.
  - Occupancy sensors: These sensors detect movement and turn off lights in unoccupied areas, reducing energy consumption.
- Adaptive lighting systems: Systems that adjust lighting in response to natural light availability and occupancy can significantly reduce energy consumption.
- Environmental and cost benefits: Transitioning to energy-efficient lighting not only saves money on energy bills, but it also reduces the environmental impact of building operations.

Smart control systems for lighting and HVAC

Smart control systems play an important role in optimising energy use in buildings.

- Automated controls: Technologies such as automated dimming, motion sensors, and programmable thermostats adjust settings to maximise energy efficiency.
  - Lighting controls: Smart lighting systems enable precise control of individual fixtures or zones. Dimmers, timers, and programmable switches improve energy efficiency.
  - HVAC controls: Advanced control systems regulate heating and cooling according to occupancy, time of day, and outside conditions. These systems ensure efficient operation while remaining comfortable.
- Responsive adjustments: These systems adapt dynamically to changes in occupancy, environmental conditions, and time of day.
- Building management integration: By centralising control of these systems within a building management system, monitoring and management become more effective, resulting in greater energy savings.

Optimised HVAC system design and maintenance

Efficient HVAC design and maintenance are critical for reducing energy consumption:

- System optimisation: Proper HVAC system sizing and zoning, combined with the use of energy-efficient components, can significantly improve energy efficiency.
- Regular maintenance of HVAC systems reduces inefficiencies and increases equipment life.
- Energy recovery ventilation: These systems use exhaust air to pre-condition incoming fresh air, reducing the demand on HVAC systems.
- **Optimised HVAC design**:
    - Proper sizing: Oversized HVAC systems waste energy and cause inefficient operation. Properly sized systems correspond to the building's load requirements.
    - Zoning: Dividing the building into zones allows for individual temperature control. Unoccupied areas require less conditioning, resulting in increased energy efficiency.
    - Variable refrigerant flow (VRF) systems: These systems adjust refrigerant flow to match the load, resulting in increased efficiency.
    - Heat recovery ventilation (HRV) systems extract heat from exhaust air and transfer it to incoming fresh air, reducing the need for additional heating or cooling.
    - Regular maintenance: Clean filters, calibrated thermostats, and well-maintained equipment ensure peak performance.

6.2.7 Impact on building energy efficiency and indoor environmental quality

Upgrading lighting and HVAC systems has a profound impact:

- Energy efficiency: Better lighting and HVAC systems directly contribute to lower energy consumption in buildings.
- Indoor environmental quality: These upgrades not only save energy but also improve occupant comfort, air quality, and productivity, resulting in more conducive indoor environments for occupant well-being.

Impact on overall building energy efficiency and indoor environmental quality:

- Implementing energy-saving strategies has a positive impact on the overall energy consumption of a building.

- Efficient lighting and HVAC systems reduce utility bills while also helping to achieve sustainability goals.
- Well-designed systems improve indoor air quality, occupant comfort, and productivity.

Water efficiency and sustainable landscaping

Water efficiency and landscaping are essential to the overall energy conservation strategy:

- Water-saving fixtures: Installing low-flow fixtures and efficient irrigation systems can help save water.
- Sustainable landscaping: Techniques such as using native plants and installing green roofs help a building's energy efficiency by increasing insulation and reducing heat island effects.

6.3 HVAC Energy Conservation Checklist

This checklist outlines a systematic approach to evaluating and optimising HVAC systems for greater energy efficiency and performance. Each item identifies a specific aspect to examine and suggests corrective actions to ensure energy efficiency.

1. Thermostat settings

- Item to check: Check thermostat settings for accuracy and suitability for the current occupancy and weather conditions.
- Corrective action: Set the thermostat to the optimal levels recommended for energy savings (20 °C (68 °F) in winter, 25.5 °C (78 °F) in summer), and update programmable settings to reflect current usage patterns.

2. Air filter maintenance

- Item to check: Check air filters for cleanliness and blockages.
- Corrective action: Clean or replace air filters on a regular basis to ensure airflow efficiency and reduce strain on the HVAC system.

3. Ductwork integrity

- Item to check: Examine the ductwork for leaks, blockages, or damage.
- Corrective action: Use duct sealant to repair leaks and remove blockages. Insulating ducts should be considered to reduce heat loss and increase efficiency.

4. Cooling and heating coils

- Item to check: Examine the cooling and heating coils for dirt and debris accumulation.
- Corrective action: Coils should be cleaned on a regular basis to ensure proper heat exchange and system efficiency.

5. System calibration

- Item to check: Check to ensure that all HVAC controls and sensors are calibrated and functional.
- Corrective action: Adjust sensors and controls as needed to ensure proper temperature and humidity management.

6. Condenser and evaporator fans

- Item to check: Check the condenser and evaporator fan blades for wear and damage.
- Corrective action: Clean fan blades to increase airflow and replace any damaged or worn fans to improve efficiency.

7. Refrigerant levels

- Item to check: Inspect refrigerant levels and detect leaks.
- Corrective action: Recharge refrigerant or repair leaks as needed to maintain peak cooling performance and efficiency.

8. Economiser function

- Item to check: Examine the operation of the economiser to see if it is bringing in sufficient amounts of outside air.
- Corrective action: Adjust or repair economiser controls to maximise the use of outside air for cooling, thereby reducing the load on mechanical cooling components.

9. Energy management system integration

- Item to check: Confirm integration and communication with the building's energy management system.
- Corrective action: Improve system integration to optimise HVAC operations using real-time data and occupancy sensors.

10. Seasonal performance

- Item to check: Evaluate how HVAC system performance and efficiency change during seasonal transitions.

- Corrective action: Perform seasonal maintenance checks and system adjustments to accommodate changing weather conditions, ensuring year-round operational efficiency.

11. Variable frequency drives (VFDs)

- Item to check: Determine whether VFDs are installed on HVAC fans and pumps.
- Corrective action: Where applicable, install VFDs to reduce motor energy consumption during low-demand periods.

12. Operational schedules

- Item to check: Check HVAC system operational schedules to ensure they are in line with building occupancy.
- Corrective action: Change schedules to reduce operation during unoccupied periods, and use programmable or smart controls to maximise energy efficiency.

13. Staff training

- Item to check: Determine whether facility personnel are trained in the proper operation and maintenance of HVAC systems.
- Corrective action: Provide staff with training and resources to help them better understand and maintain HVAC efficiency.

References

[1] IEA (2023), Energy Efficiency 2023, IEA, Paris https://www.iea.org/reports/energy-efficiency-2023,Licence:CCBY4.0. (Accessed: 08 May 2024)

[2] IEA (2023), World Energy Outlook 2023, IEA, Paris https://www.iea.org/reports/world-energy-outlook-2023,Licence:CCBY4.0(report);CC BYNCSA4.0(AnnexA). (Accessed: 08 May 2024)

[3] Barney L. Cape hart, Wayne C. Turner (2016), Guide to Energy Management, Fairmont Press, ISBN: 1420084895

[4] Thomson, E. (2023) Half-price energy bills if you live near a wind farm. Here's some incentives available in different countries to accelerate the renewable energy transition, World Economic Forum. Available at: https://www.weforum.org/agenda/2023/05/renewable-energy-incentives-households-countries/. (Accessed: 08 May 2024)

[5] Energy end-use data collection methodologies and the emerging role of digital technologies – Analysis. IEA. Available at: https://www.iea.org/reports/energy-end-use-data-collection-methodologies-and-the-emerging-role-of-digital-technologies.

[6] Treaties (2016) www.iaea.org. Available at: https://www.iaea.org/resources/legal/treaties. (Accessed: 08 May 2024)

[7] Practical ways to apply data analytics and AI for energy management and emission control in the steel and cement industries (2024) Industrial Software. Available at:https://new.abb.com/industrial-software/digital/energy-managment/practical-ways-to-apply-data-analytics-and-ai-for-energy-management-and-emission-control-in-the-steel-and-cement-industries. (Accessed: 08 May 2024)

[8] Conserving energy in transportation | Global Strategies & Solutions | The Encyclopaedia of World Problems (2016) Uia.org. Available at: http://encyclopedia.uia.org/en/strategy/199528 (Accessed: 8 May 2024).

[9] Making the energy sector more resilient to climate change – Analysis, IEA. Available at: https://www.iea.org/reports/making-the-energy-sector-more-resilient-to-climate-change. (Accessed: 08 May 2024)

[10] Six key steps to become an energy efficiency expert | IDB Invest, www.idbinvest.org. Available at: https://www.idbinvest.org/en/blog/energy/six-key-steps-become-energy-efficiency-expert (Accessed: 8 May 2024).

7

Energy Management

Chapter Preview

This chapter starts by defining energy management and the need for it. We discuss how to design, start, and manage an energy management program. We also explore energy accounting, energy efficiency in industrial processes, energy management systems, demand side management, and thermal energy storage.

Definition

'The efficient and effective use of energy to maximise profits (lower costs) and improve competitiveness'.

This broad definition encompasses a variety of operations, including services, product and equipment design, and product shipment. Waste minimisation and disposal also provide numerous energy management opportunities.

The primary goal of energy management is to increase profits or reduce costs. Some desirable sub-objectives of energy management programmes are:

1. improving energy efficiency and lowering energy consumption, which reduces costs;
2. reducing greenhouse gas emissions while improving air quality;
3. fostering effective communication on energy issues;
4. creating and implementing effective monitoring, reporting, and management strategies for efficient energy use;
5. finding new and better ways to increase the returns on energy investments through research and development;
6. creating interest in and commitment to the energy management programme among all employees;
7. reducing the effects of curtailments*, brownouts, or other interruptions in energy supplies.

**Curtailments occur when a major supplier of an energy source is forced to reduce shipments or allocations (sometimes significantly) due to severe* *weather and/or distribution issues. For example, natural gas is frequently sold to industry at a low cost but with intermittent availability. Even though curtailments are infrequent, the costs associated with them are so high—sometimes a complete shutdown is required—that management must be vigilant to mitigate the negative effects. There are several methods for doing this, but the most common one is to store and use a secondary or standby fuel.*

Energy conservation is undoubtedly important in energy management, but it is not the only consideration. Other factors to consider include curtailment planning, load shedding, and power factor improvement.

7.1 The Need for Energy Management

7.1.1 Economics

In any free enterprise system, any new activity can be justified if it is cost-effective, which means that the net result must show a profit increase or cost reduction greater than the activity's cost. Energy management has repeatedly proven to be cost effective.

When an aggressive energy management programme is launched, energy costs are typically reduced by 5%–15% with little to no capital expenditure. Savings of 30% are common, with savings of 50%, 60%, and even 70% obtained. These savings are all the result of retrofitting activities. New energy-efficient buildings can use 20% of the energy that existing buildings do (resulting in an 80% savings). Indeed, for the majority of manufacturing, industrial, and commercial organisations, energy management is one of the most promising profit improvement-cost reduction programmes available today.

7.1.2 National and global good

Energy management programmes are extremely important today. One important reason is that energy management helps the country address some of its most pressing issues. Increasing population, energy/oil imports, and reliance on imported oil.

Energy use has also caused significant environmental, economic, and industrial competitiveness issues. Reducing energy consumption can help to mitigate these issues by:

i. Acid rain has been reduced primarily through national and regional environmental policies around the world.
ii. Limiting global climate change. Carbon dioxide, the primary contributor to potential global climate change, is produced by the combustion of

fossil fuels, which are primarily used for transportation and energy purposes. Reducing fossil energy consumption through energy efficiency improvements and the use of renewable energy is the quickest, most effective, and cost-effective way to reduce greenhouse gas emissions and improve air quality, especially in densely populated areas.

iii. Reduce ozone depletion. The Montréal Protocol of 1987, and subsequent updates, is one of the world's most successful environmental protection treaties. The Protocol establishes a mandatory timetable for the phase-out of ozone-depleting substances.

iv. Improving national security. Oil imports have a direct impact on our country's energy security and balance of payments. These oil imports must be reduced to ensure a stable political and economic future.

v. Increased a country's competitiveness.

Some possible solutions to the energy problem.

i. Many people believe that coal holds the answer. However, coal combustion emits sulphur dioxide and carbon dioxide, which cause acid rain and potential global climate change. Research and development for 'clean coal' technology is currently underway.

ii. Synfuels necessitate strip mining, have high costs, and place enormous demands on water in arid areas. Several electric utilities are currently demonstrating on-site coal gasification plants alongside gas-fired combined-cycle power plants. However, it remains to be seen whether these units can be constructed and operated in a cost-effective and environmentally friendly manner.

iii. Solar electricity, whether generated by photovoltaics or thermal processes, remains more expensive than conventional sources and requires a large amount of land. Both areas are seeing technological advancements, as well as cost reductions. These approaches may eventually become cost-effective.

iv. Biomass energy is also expensive, and any type of monoculture would necessitate a large plot of land. Some people fear that forests will be destroyed completely. At best, biomass can meet only a small percentage of our total needs without significant problems.

v. Wind energy is only feasible in certain geographical areas where wind velocity is consistently high, and there are some noise and aesthetic concerns. However, the cost of wind generation systems has decreased, making them cost-effective in windy areas. Operating costs are also low.

vi. Fuel cells and hydrogen are attractive due to their ability to generate electricity cleanly from hydrogen and oxygen. However, hydrogen is not the primary source of energy. It is derived from other forms of energy; today, the majority of hydrogen is produced by steam reforming natural gas. Natural gas is a fossil fuel, so the carbon dioxide released during the reformation process contributes to the greenhouse effect. Only when hydrogen is produced in a cost-effective manner from renewable energy sources does it have significant value as a fuel source for fuel cells.

vii. Alcohol production from agricultural products raises perplexing questions about using food products as energy when large parts of the world are hungry. Newer processes for producing alcohol from wood waste are still being tested, and they could significantly improve this limitation.

viii. Fission has well-known waste disposal and safety issues, as well as a short-time span using current technology. Without breeder reactors or nuclear fuel reprocessing, we will soon run out of fuel, but breeder reactors significantly increase plutonium production—a raw material for nuclear bombs. Nuclear fuel reprocessing could provide many years' worth of fuel by recycling partially used fuel that is currently stored.

ix. Fusion appears to be everyone's hope for the future, but many argue that we do not know enough about the subject to predict its problems. When available commercially, fusion may present its own set of environmental and economic issues.

Energy management has repeatedly demonstrated its ability to significantly reduce energy costs and consumption through increased energy efficiency. This saved energy can be applied elsewhere. In fact, energy generated through energy management activities has almost always proven to be the most cost-effective source of 'new' energy. Furthermore, energy management activities are gentler on the environment than large-scale energy production, and they do result in less consumption of scarce and valuable resources. Thus, while energy management cannot solve all of the nation's problems, it may be able to reduce our environmental impact and give us more time to develop new energy sources.

7.2 Designing an Energy Management Programme

7.2.1 Management commitment

Top management's commitment to the programme is the single most important factor in the successful implementation and operation of an energy management programme. Without this commitment, the programme will most likely fail to meet its objectives. Thus, the role of the energy manager is critical in ensuring management's commitment to the programme.

When designing an energy management programme, two situations are equally likely to occur. First, management has decided that energy management is necessary and wishes to implement a programme. This puts you – the energy manager – in response mode. In the second scenario, you, as an employee, have decided to persuade management that the programme is necessary, so you are in aggressive mode. Obviously, the response mode is the best situation because much of your sales effort is unnecessary; however, many energy management programmes have been launched using the aggressive mode.

Response mode

In a typical response mode scenario, management has observed rapidly rising energy prices and/or curtailments, has learned about the results of other energy management programmes, and has taken action to begin the programme. In this case, the management commitment is already in place; all that remains is to cultivate that commitment on a regular basis and ensure that it is visible to all programme participants.

Aggressive mode

In the aggressive mode, you, the employee, are aware that energy costs are skyrocketing and that sources are less secure. You may have taken an energy management course, attended a professional conference, or read related papers. At any rate, you are convinced that the company requires an energy management programme. All that remains is to persuade management and secure their commitment. The best way to persuade management is through facts and statistics. Graphs can sometimes be the most startling way to present facts. Follow this data with quotes from other companies' programmes to demonstrate that these goals are achievable.

7.2.2 Energy management coordinator/energy manager

To develop and maintain the vitality of the energy management programme, a company must assign a single person to coordinate it. Assume that no one's job description includes energy management. In that case, management is likely to discover that energy management efforts are given less priority than other job duties. As a result, little to no progress may be made.

The energy management coordinator (EMC) should be strong, dynamic, goal-oriented, and an effective manager. Most importantly, management should provide that individual with resources, including staff. The energy management coordinator should report as high as possible in the organisation while maintaining line orientation.

7.2.3 Backup talent

A comprehensive study of the plant's steam production, distribution, usage, and condensate return system may necessitate the use of multiple engineering disciplines. As a result, most effective energy management programmes include an energy management committee. The technical and steering sub-committees are often desirable.

The technical committee

The technical committee is usually made up of several people with a strong technical background in their field. This committee may include representatives from chemical, industrial, electrical, civil, and mechanical engineering, among other disciplines. Their role is to provide technical assistance to the coordinator and plant-level employees. For example, the committee can stay up to date on emerging technology and conduct research into potential applications across the company. The results can then be narrowed down.

While the energy management coordinator may be a full-time position, the technical committee is likely to work part-time and be called upon as needed.

The steering committee

The steering committee serves a completely different purpose than the technical committee. It guides the energy management program's activities and facilitates communication at all levels of the organisation. The steering committee also ensures that all plant employees are aware of the programme. The steering committee members are usually chosen to represent all major areas of the company. Steering committee members should be chosen based on their broad interests and genuine desire to help solve energy problems.

7.2.4 Cost allocation

One of the most difficult challenges for the energy manager is attempting to reduce a facility's energy costs when they are accounted for as part of general overhead. In that case, individual managers and supervisors do not see themselves as responsible for controlling energy costs. This is because they do not see a direct benefit to lowering costs that are part of the overall company overhead. The best solution to this problem is for top management to allocate energy costs to specific 'cost centres' within the company or facility. When energy costs are charged to production centres in the same way that materials and labour are, managers have a direct incentive to control those costs because doing so improves the production centre's overall cost-effectiveness. That is, they will have a stake in the game.

7.2.5 Reporting and monitoring

The energy management coordinator and steering committee must be aware of the plant's 'pulse of energy consumption'. This is best accomplished by implementing an effective and efficient energy reporting system. An energy reporting system measures and compares energy consumption against company goals or standards. Ideally, this should be done for each operation or production cost centre in the plant, but most facilities lack the necessary metering equipment. Many plants only measure energy consumption at a single point – where the various sources enter the plant. Most plants, however, are attempting to address this by installing additional metering devices as opportunities arise. As always, the reporting scheme should be reviewed on a regular basis to ensure that only necessary material is generated, that all required data is available, and that the system is efficient and effective.

7.2.6 Training

Most energy management coordinators believe that substantial training is required. Training does not happen overnight, and it is never 'completed'.

Table 7.1 Energy management training.

Personnel involved	Type of necessary training	Source of required training
Technical committee	Sensitivity to energy management	In house (with outside help?)
	Technology developments	Professional societies, universities, consulting groups, journals
Steering committee	Sensitivity to energy management Other industries' experience	Trade journals, energy sharing groups, consultants
Plant-wide	Sensitivity to energy management What's expected, goals to be obtained, etc.	In house

Changes occur among energy management personnel and employees at all levels, as well as new technology and manufacturing methods. All of these result in training or retraining. The energy management coordinator is responsible for this training (Table 7.1).

7.3 Starting an Energy Management Programme

Several factors contribute to the successful implementation of an energy management programme. They include:

1. visibility of the program's launch;
2. management's commitment to the programme has been demonstrated;
3. selecting an excellent initial energy management project.

7.3.1 Visibility of the programme's launch

To be effective, an energy management programme must have the support of those involved. Obtaining this support is often a difficult task, so careful planning is required. The citizens must:

1. Learn why the programme exists and what its goals are.
2. Consider how the programme will impact their jobs and income.
3. Understand that the programme has full management support.
4. Understand what is expected of them.

Management and the energy management coordinator collaborate on communicating this information to employees.

7.3.2 Demonstration of management commitment

Management's commitment to the programme is critical, and it must be visible to all employees if the programme is to reach its full potential. Management's involvement in the program's launch demonstrates this commitment, but it should also be highlighted in other ways. For example:

1. Recognise and reward those who participate. Recognition is a powerful motivator for most employees. An employee who has been a staunch supporter of the programme deserves recognition.
2. Reinforce commitment. Management must recognise that employees are constantly watching them.
3. Support cost-effective proposals. All businesses face capital budgeting challenges of varying severity, and unfortunately, energy projects are not

given the same priority as front-line items like equipment acquisition. However, management must recognise that rejecting the energy management team's proposals while accepting others with lower economic attractiveness is a surefire way to kill enthusiasm.

7.3.3 Early project selection

The energy management programme is off to a rocky start. Most employees are concerned that their heat will be reduced, their air conditioning will be turned off, and their lighting will be dimmed. If any of these actions take place, it is no surprise that employee support is waning. These things may happen in the future, but would not it be better to start with less controversial projects?

An early failure can be detrimental to the programme, if not disastrous. As a result, the astute energy management coordinator will 'stack the deck' for his or her first set of projects. These projects should have a quick payback, a high chance of success, and minimal negative consequences.

7.4 Management of the Programme

7.4.1 Establishing objectives in an energy management program

To ensure a successful energy management programme, management should promote creativity rather than discourage it. Goals must be established, and these goals should be challenging but attainable, measurable, and specific. They must also specify a deadline for completion. Once management and the energy management coordinator have agreed on the goals and established a good monitoring or reporting system, the coordinator should be left alone to complete his or her duties.

The following list provides some examples of such goals:

- Total energy per unit of production will decrease by 10% the first year and another 5% the second.
- By the end of the first year, each plant in the division will have implemented an active energy management programme.

7.5 Energy Accounting

Energy accounting is a system for keeping track of energy consumption and expenses. A basic energy accounting system consists of three components:

energy use monitoring, an energy use record, and a performance measure. The performance metric can range from a simple index of Btu/m^2 or Btu/unit of production to a sophisticated standard cost system with variance reporting. Metering is required for all energy accounting purposes. To monitor energy flow through a cost centre, it is necessary to measure both incoming and outgoing energy, regardless of size. The lack of necessary metres is probably the most significant single impediment to the widespread use of energy accounting systems.

7.5.1 Levels of energy accounting

Energy accounting systems, like financial accounting, vary greatly in sophistication and detail from one company to the next. There is a strong correlation between the sophistication levels of financial accounting systems and those of energy accounting systems (Table 7.2).

7.5.2 Performance measures

7.5.2.1 Energy utilisation index

The energy utilisation index is a basic measure of a facility's energy performance (EUI). This is a statement of the annual Btu energy consumption per square foot or metre of conditioned space. To calculate the EUI, all energy consumed in the facility must be identified, the total Btu content calculated, and the total number of square metres of conditioned space determined. The EUI is then calculated as the ratio of total Btu consumed to total square feet of conditioned space.

Table 7.2 Comparison between financial and energy accounting.

Financial	Energy
General accounting	Effective metering, development of reports, calculation of energy efficiency indices
Cost accounting	Calculation of energy flows and efficiency of utilisation for various cost centres; requires substantial metering
Standard cost accounting historical standards	Effective cost centre metering of energy and comparison to historical data; complete with variance reports and calculation of reasons for variation
Standard cost accounting engineered standards	Same as above except that standards for energy consumption are determined through accurate engineering models.

Example 7.1

Consider a building with 100,000 square feet of floor space. It uses 1.76 million kWh and 6.5 million cubic feet of natural gas in one year. Find the energy utilisation index (EUI) for this facility.

Solution: Each kWh contains 3412 Btu and each cubic foot of gas contains about 1000 Btu. Therefore, the total annual energy use is:

Total energy use

$$\begin{aligned} &= \left(1.76 \times 10^6 \text{ kWh}\right) \times (3412 \text{ Btu/kWh}) \\ &\quad + \left(6.5 \times 10^6 \text{ ft}^3\right) \times \left(\frac{1000 \text{ Btu}}{\text{ft}^3}\right) \\ &= 6.0 \times 10^9 + 6.5 \times 10^9 \\ &= 1.25 \times 10^{10} \text{ Btu/yr.} \end{aligned}$$

Dividing the total energy use by 10^5 ft^2 gives the EUI:

$$\begin{aligned} \text{EUI} &= \left(1.25 \times 10^{10} \text{ Btu/yr}\right)/\left(10^5 \text{ ft}^2\right) \\ &= 125,000 \text{ Btu/ft}^2\text{/yr.} \end{aligned}$$

7.5.2.2 Energy cost index

Another useful performance metric is the energy cost index, or ECI. This is a statement of the annual cost of energy used per square foot of conditioned space. To calculate the ECI, all energy consumed in the facility must be identified, the total cost of that energy calculated, and the total number of square feet of conditioned space determined. The ECI is calculated by dividing a facility's total annual energy cost by its total square feet of conditioned floor space.

Example 7.2

Consider the building in Example 7.1. The annual cost for electric energy is \$115, 000 and the annual cost for natural gas is \$32,500. Find the energy cost index (ECI) for this facility.

Solution: The ECI is the total annual energy cost divided by the total number of conditioned square feet of floor space.

$$\text{Total energy cost} = 115,000 + 32,500 = 147,000\text{/yr.}$$

Dividing this total energy cost by 100,000 square feet of space gives :

$$\text{ECI} = \frac{\frac{147{,}500}{\text{yr}}}{100,000 \text{ ft}^2} = \$\frac{1.48}{\text{ft}^2}\text{/yr.}$$

7.5.2.3 One-shot productivity measures

The energy utilisation index is plotted over time, with trends noted. Significant differences from the same period in the previous year should be noted and explained. This measure is frequently used to justify or demonstrate the effectiveness of energy management activities.

7.6 Energy Efficiency in Industrial Processes

The industrial sector faces unique energy management challenges while also offering significant conservation opportunities. Energy efficiency in industrial processes is critical not only for cost savings, but also for reducing environmental impact.

Optimising machinery and equipment efficiency

Efficient machinery and equipment are foundational to energy conservation in industrial processes:

- Maintenance and retrofitting: Regular maintenance ensures that machinery functions properly. Retrofitting older equipment with energy-efficient components can result in significant energy savings.
- Advanced technology: Investing in newer, more energy-efficient machinery yields long-term savings. Automation and control systems improve efficiency by optimising operational parameters.
- Energy management systems: Using advanced energy management systems can lead to more efficient energy use, less waste, and increased overall efficiency.

Waste heat recovery and utilisation

Waste heat recovery is a vital aspect of industrial energy efficiency:

- Recovery technologies include heat exchangers and cogeneration systems, which capture and reuse waste heat from industrial processes.
- Application examples: These technologies can benefit industries that generate a lot of waste heat, such as steel, cement, and chemical manufacturing.
- Energy savings: Using waste heat can dramatically reduce the amount of energy required for heating and power, resulting in significant energy conservation.

Process redesign for energy efficiency

Redesigning industrial processes can lead to significant energy efficiency gains:

- Streamlining production: Analysing and streamlining production processes can reduce energy consumption while maintaining output.
- Lean manufacturing: Adopting lean manufacturing principles reduces waste, including energy waste, while increasing overall efficiency.
- Innovative production techniques: By implementing new, more energy-efficient production techniques, industries can change their energy consumption patterns.

Systematic approach to industrial energy management

A systematic approach to energy management is critical for identifying and capitalising on energy-saving opportunities:

- Energy audits: Conducting thorough energy audits allows you to identify areas where you can save energy.
- Continuous monitoring: Ongoing energy usage monitoring enables the timely detection and correction of inefficiencies.
- Performance metrics: Establishing and tracking energy performance metrics allows you to evaluate the effectiveness of energy conservation measures.

Challenges and opportunities

Industrial settings present unique challenges and opportunities for energy conservation. Here are some key aspects:

1. **Machinery efficiency optimisation**:
    - Challenge: Industrial machinery consumes a lot of energy. It is critical to ensure optimal performance while using the least amount of energy possible.
    - Opportunity: Regular maintenance, proper calibration, and the use of advanced control systems can increase machinery efficiency. Variable speed drives, smart sensors, and predictive maintenance all play important roles.
2. **Waste heat recovery**:
    - Challenge: Heat generated during industrial processes is frequently unused.

- Opportunity: Waste heat recovery systems can capture and reuse this energy. Effective technologies include heat exchangers, cogeneration, and organic Rankine cycles.

3. **Process redesign**:
    - Challenge: Existing processes may be energy-intensive or inefficient.
    - Opportunity: Rethinking process design can result in significant energy savings. Consider alternative materials, streamlined workflows, and modular manufacturing methods.

Importance of systematic energy management

A haphazard approach won't suffice. Industries must adopt a systematic strategy:

1. **Energy audits**:
    - Regular assessments identify energy consumption patterns, inefficiencies, and opportunities for improvement.
    - Audits help guide decision-making and prioritise energy-saving initiatives.
2. **Energy performance indicators (EnPIs)**:
    - Establishing EnPIs allows you to track progress and benchmark energy efficiency.
    - Metrics such as energy intensity (energy consumed per unit of production) provide useful information.
3. **Employee training and engagement**:
    - Educate employees about energy-saving practices.
    - Engaged employees help to sustain energy conservation efforts.
4. **Investment in technology**:
    - Embrace automation, data analytics, and energy management software.
    - Real-time monitoring and control improves efficiency.
5. **Policy and regulatory compliance**:
    - Adhere to energy regulations and standards.
    - Compliance ensures energy efficiency and environmental stewardship.

7.6.1 Real-world case studies

1. **Saving energy in industrial companies**:
   - This comprehensive report, 'Saving Energy in Industrial Companies: Case Studies of Energy Efficiency Programs in Large U.S. Industrial Corporations and the Role of Ratepayer-Funded Support' focuses on large U.S. industrial corporations and their energy efficiency programs. It highlights the role of ratepayer-funded support and provides case studies of successful energy-saving initiatives.
   - Key topics covered include:
     - **Untapped energy efficiency potential**: Identifying areas where energy efficiency gains can be made.
     - **Corporate structure and culture**: Overcoming challenges within the organisational framework.
     - **Effective organisation arrangements**: Strategies for successful energy efficiency programs.
     - **Ratepayer-funded programmes**: How they contribute to overcoming challenges.
   - The report offers insights from real-world case studies, making it a valuable resource for professionals in the field.
   - Source: https://www.energy.gov/sites/default/files/2021-07/saving_energy_industrials.pdf
2. **Case studies by system**:
   - The U.S. Department of Energy (DOE) provides case studies documenting energy savings achieved by large manufacturing companies using various tools and best practices.
   - These case studies cover systems such as:
     - Steam
     - Process heating
     - Compressed air
     - Motor
     - Pump
     - Fan
     - Plant wide

- Each case study showcases successful energy management practices and their impact.
- Source: https://www.energy.gov/eere/iedo/case-studies-system

3. **Industrial energy efficiency project (IEE) case studies**:
    - The IEE Project Phase II (2016–2022) features case studies that demonstrate effective energy management systems (EnMS) in industrial settings.
    - Notable examples include:
        - **Auto-X**: An EnMS case study.
        - **Belgotex floors**: A successful implementation of energy management practices.
    - These case studies provide practical insights for professionals aiming to optimise energy usage.
    - Source: https://www.energy.gov/sites/default/files/2021-07/saving_energy_industrials.pdf
4. **Smart energy management for industrials**:
    - According to Deloitte's 2020 Resources Study, industrial respondents cited cost reduction as the primary driver for implementing energy management programs.
    - Smart energy management becomes crucial for industrial companies as electricity rates are expected to rise.
    - Source: https://www2.deloitte.com/us/en/insights/industry/power-and-utilitiesutilities/smart-energy-management.html

7.7 Energy Management Systems

Energy management systems (EMS) are critical tools for monitoring, controlling, and optimising the distribution and use of energy in buildings, factories, and other facilities. An energy management system (EMS) aims to improve energy efficiency, reduce energy consumption, and lower operational costs while maintaining comfort and productivity.

7.7.1 Components of energy management systems

1. Sensors and metres collect data on a variety of parameters, including temperature, humidity, and electricity usage. They are critical to providing the real-time data required for energy monitoring.

2. Controllers: Controllers automate system operations based on sensor data. They play an important role in maintaining energy efficiency by automatically adjusting settings to meet energy-saving standards.
3. User interface (UI): This allows users to interact with the EMS, view data, and manually override automatic controls as needed. UIs are typically designed to be user-friendly and accessible across a variety of devices, including smartphones and computers.
4. Data analytics: Advanced analytics are used to identify trends, forecast energy consumption, and recommend areas for improvement. These insights are critical for continuous energy optimisation.

Types of energy management systems

1. Home energy management systems (HEMS): These systems are designed for residential use, allowing homeowners to monitor and control their energy consumption using smart thermostats and automated lighting.
2. Building energy management systems (BEMS): These systems, which are used in commercial buildings, manage various building systems such as HVAC, lighting, and power to ensure that energy is used efficiently.
3. Industrial energy management systems (IEMS): These are designed for industrial applications, with a focus on optimising manufacturing processes and lowering energy costs in production environments.

Benefits of energy management systems

- Cost savings: One of the most significant advantages of EMS is the reduction in energy expenses. Households and businesses can save money on their energy bills by optimising their energy use.
- Enhanced sustainability: EMS helps to reduce carbon footprints and promote sustainability by reducing energy waste and increasing energy efficiency.
- Improved system performance: EMS-facilitated regular monitoring and maintenance improves the performance and longevity of building systems and appliances.
- Compliance and reporting: Many regions have regulations that require energy monitoring and reporting. An EMS can automate these tasks, ensuring that they meet local or international standards.

Trends and future technologies

- Integration with IoT: Integrating EMS with the Internet of things (IoT) enables more connected, smarter systems that can better control energy consumption across multiple devices.
- Machine learning and artificial intelligence: These technologies are increasingly being used to predict energy usage patterns and automate system adjustments, thereby optimising energy consumption.
- Energy storage systems: As renewable energy usage grows, EMS technologies are increasingly focusing on the efficient integration and management of energy storage systems.

7.7.2 Energy management systems checklist for energy conservation

This checklist is intended to assist facilities in optimising their energy management systems (EMS) to achieve maximum energy efficiency and conservation. It covers various EMS components and functions, including check guidelines and corrective actions.

1. **System calibration and sensors**
    - **Item to check:** Verify calibration of all sensors (temperature, humidity, occupancy, etc.).
    - **Corrective action:** If inaccuracies are detected, recalibrate sensors or replace faulty sensors to ensure accurate data collection and system responsiveness.
2. **User interface and control settings**
    - **Item to check:** Check the ease of use and accessibility of the EMS user interface. Examine the control settings for heating, cooling, lighting, and other systems.
    - **Corrective action:** Improve the user interface for a better user experience and set settings to optimal levels for energy conservation.
3. **Software and firmware updates**
    - **Item to check:** Make sure the EMS software and associated firmware are up to date.
    - **Corrective action:** Install the most recent software and firmware updates to improve system functionality and introduce new energy-saving features.

4. **System integration and communication**
    - **Item to check:** Ensure that all EMS components (HVAC, lighting, and security) are integrated and communicate effectively.
    - **Corrective action:** Troubleshoot and resolve any integration issues. Improve system communication to ensure smooth operation across all connected devices.
5. **Energy consumption data**
    - **Item to check:** Examine energy consumption data for anomalies or patterns that suggest inefficiency.
    - **Corrective action:** Adjust system parameters to address inefficiencies and implement specific energy-saving measures.
6. **Scheduling and timers**
    - **Item to check:** Review HVAC and lighting system scheduling settings to ensure they match occupancy and usage patterns.
    - **Corrective action:** Reconfigure schedules and timer settings to avoid wasting energy during unoccupied periods.
7. **Demand response and peak load management**
    - **Item to check:** Assess the efficacy of demand response systems and peak load management approaches.
    - **Corrective action:** Optimise or implement new demand response strategies to reduce energy consumption during peak periods while also taking advantage of utility incentives.
8. **HVAC system efficiency**
    - **Item to check:** Inspect HVAC systems for proper operation and maintenance status.
    - **Corrective action:** Perform maintenance or upgrades as needed; modify HVAC operations to increase efficiency and reduce energy consumption.
9. **Lighting efficiency**
    - **Item to check:** Determine the efficiency of lighting systems, including fixture types, bulb efficiency, and control systems.
    - **Corrective action:** Upgrade to more efficient lighting solutions (such as LEDs); and improve control systems with motion sensors and dimming capabilities.

10. **Renewable energy integration**
    - **Item to check:** Evaluate the integration and performance of renewable energy sources (solar, wind, etc.) with the EMS.
    - **Corrective action:** Improve integration techniques and system configuration to make the best use of renewable energy within the facility.
11. **Reporting and monitoring tools**
    - **Item to check:** Check the EMS's reporting and monitoring tools to ensure they are adequate for real-time energy tracking and historical analysis.
    - **Corrective action:** Upgrade or implement advanced monitoring and reporting tools to improve energy efficiency and savings.
12. **Training and user awareness**
    - **Item to check:** Evaluate staff training and awareness of the EMS's capabilities.
    - **Corrective action:** Conduct regular training sessions to ensure that employees are knowledgeable about operating the EMS efficiently and aware of energy conservation practices.

7.8 Demand Side Management

7.8.1 Introduction

Definition

'Demand-side management is the planning, implementation, and monitoring of utility activities aimed at influencing customer electricity consumption in ways that result in desired changes in the utility's load shape, i.e., changes in the time pattern and magnitude of a utility's load. Load management, new uses, strategic conservation, electrification, customer generation, and market share adjustments are examples of utility programmes classified as demand-side management'.

Demand-side management embraces the following critical components of energy planning:

1. Demand-side management will have an impact on customer behaviour. Demand-side management refers to any programme that aims to influence a customer's energy use.

2. Demand-side management must meet specific objectives. To achieve the desired load-shape change, the programme must contribute to the achievement of specific objectives, such as lower average rates, increased customer satisfaction, meeting reliability targets, and so on.
3. Demand-side management will be compared to other types of management. The concept also requires that certain demand-side management programmes advance these goals at least as much as non-demand-side management alternatives such as generating units, purchased power, or supply-side storage devices. In other words, demand-side management alternatives must be weighed against supply-side options. At this point in the evaluation, demand-side management is integrated into the resource planning process.
4. Demand-side management predicts how customers will respond. Demand-side management is a pragmatic approach. Normative programmes (we should do this) do not produce the desired results; positive efforts are required. Demand-side management identifies how customers will respond, rather than how they should.
5. Load shape influences demand-side management value. Finally, this definition of demand-side management emphasises the load shape. This implies an evaluation process that looks at the value of programmes based on how they affect costs and benefits throughout the day, week, month, and year.

7.8.2 Demand-side management and integrated resource planning

A critical component of the demand-side management process is the consistent evaluation of demand-side to supply-side alternatives, and vice versa. This approach is known as integrated resource planning. For demand-side management to be a viable resource option, it must compete with traditional supply-side alternatives.

7.8.3 Demand-side management programmes

Since the early 1980s, when demand-side management was introduced, a variety of programmes have been implemented.

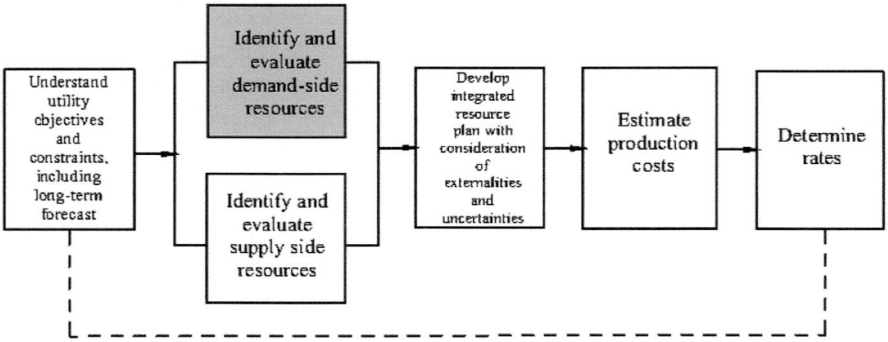

Figure 7.1 How demand-side management fits into integrated resource planning.

7.8.3.1 Elements of the demand-side management planning framework

1. **Set objectives**: The first step in demand-side management planning is to define overall organisational goals. These strategic objectives are quite broad and generally include examples such as conserving energy resources, reducing peak demand (thus avoiding the need to build new power plants), lowering greenhouse gas emissions, reducing reliance on foreign imports, improving cash flow, increasing earnings, and improving customer and employee relations. In this stage of the formal planning process, the planner must operationalise broad objectives in order to guide policymakers to specific actions. Demand-side management alternatives should be examined and evaluated at this operational level, also known as tactical. For example, examining capital investment requirements may reveal periods of high investment demand. A demand-side management programme can postpone the need for new construction, lowering investment needs and stabilising the financial future of an energy company, utility, and its state or country.
2. **Identify alternatives**: The second step is to identify alternatives. The first dimension of this step entails identifying suitable end-uses whose peak load and energy consumption characteristics broadly match the load-shape objectives established in the previous step (Table 7.3). Generally, each end-use (e.g. residential space heating, commercial lighting) has predictable demand or load patterns. One factor used to choose an end-use for demand-side management is its ability to accommodate load pattern modification. The second dimension of

Table 7.3 Six generic load-shape objectives can be considered during demand-side management planning

Peak clipping: 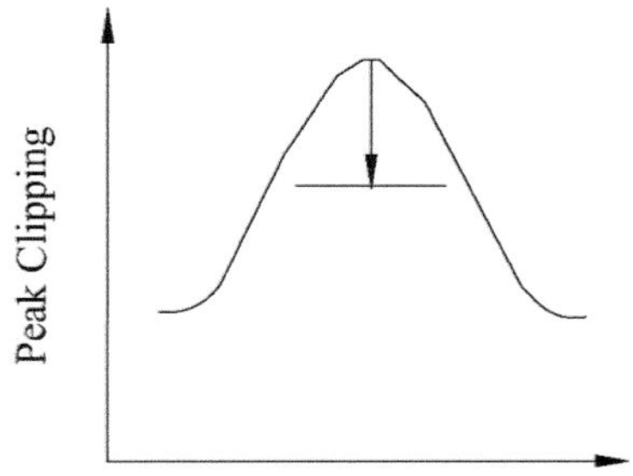	This aims to reduce electricity demand during peak hours, usually in the afternoon or evening when usage is highest. This can help to reduce the need for costly and polluting peak-generation plants.
Valley filling: 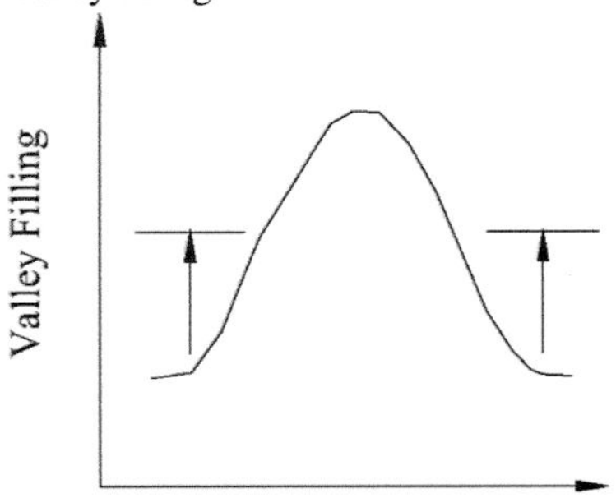	This focuses on increasing electricity demand during off-peak hours, usually overnight. This can help to use excess generation capacity and potentially reduce overall electricity costs.
Load shifting: 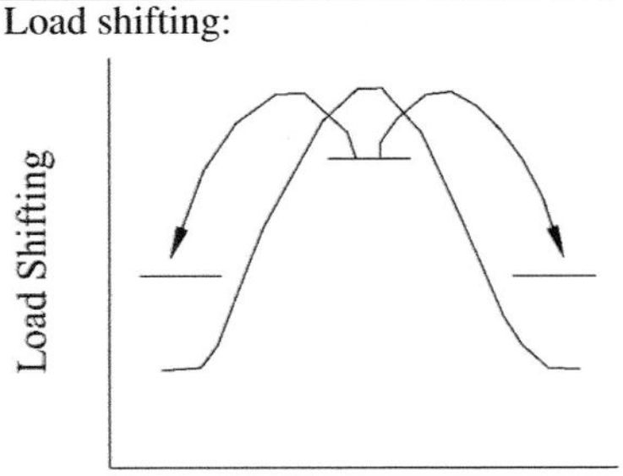	This involves shifting electricity demand from peak to off-peak hours. This can be accomplished by encouraging customers to use appliances such as dishwashers or washing machines during off-peak hours.
Strategic conservation: 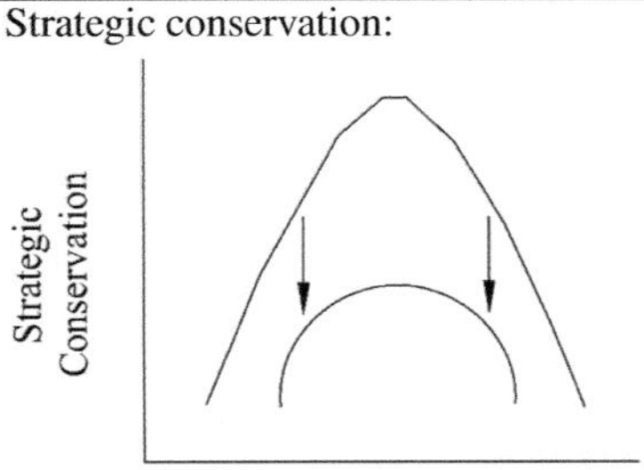	This goal seeks to permanently reduce total electricity consumption through efficiency improvements and behavioural changes.

Table 7.3 *Continued.*

Strategic load growth: 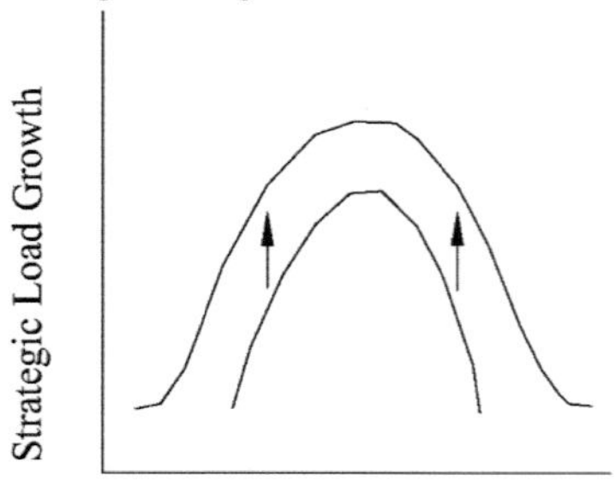	In some cases, it may be beneficial to encourage controlled growth in electricity demand, especially if it involves a shift to cleaner energy sources.
Flexible load shape: 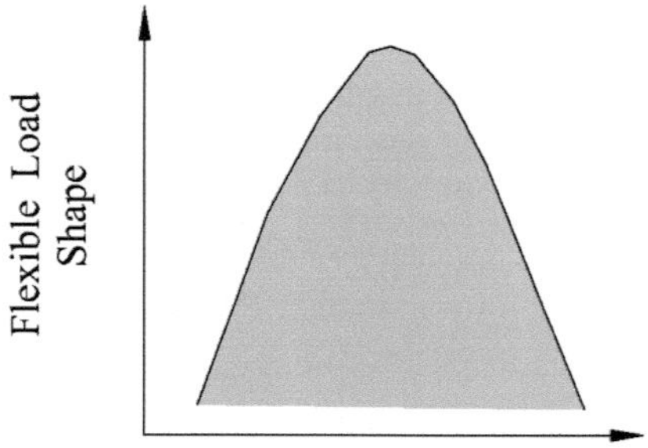	This objective aims to create a more responsive electricity demand profile. This allows utilities to better integrate renewable energy sources like solar and wind power, which can be variable in their output.

demand-side management alternatives entails selecting appropriate technology alternatives for each intended end-use. This process should consider the technology's suitability for meeting the load-shape goal.

3. **Evaluate and select programme(s)**: The third step involves balancing customer and supplier considerations, as well as cost/benefit analyses, to determine the best demand-side management options. Although customers and suppliers work independently to change demand patterns, demand-side management implies a supplier–customer relationship that yields mutually beneficial results. To achieve that mutual benefit, suppliers must carefully consider factors such as how the activity will affect demand patterns and amounts (load shape), the methods available for obtaining customer participation, and the likely magnitudes of costs and benefits to both supplier and customer before proceeding with implementation.
4. **Implement programme(s)**: The fourth step is to implement the programme(s), which occurs in stages. As a first step, a high-level demand-side management project team should be formed, with representatives

from various departments and organisations, as well as overall control and responsibility for the implementation process. Implementers must provide clear directives to the project team, such as a written scope of responsibility, project team goals, and timeline.

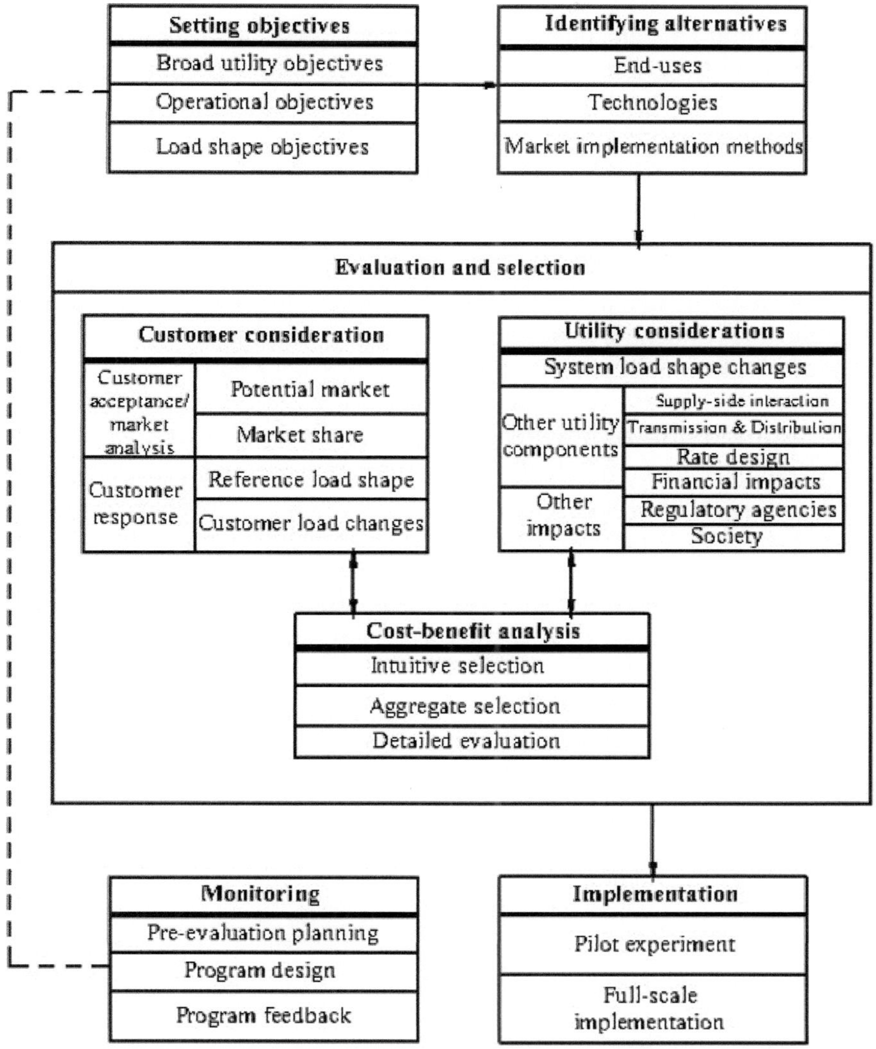

Figure 7.2 Elements of the demand-side management planning framework.

5. **Monitor programme(s)**: The fifth step involves monitoring the programme(s). The ultimate goal of the monitoring process is to detect deviations from expected performance and improve current and planned demand-side management programmes. Monitoring and evaluation processes can also serve as a primary source of information on customer behaviour and system impacts, promote advanced planning and organisation within a demand-side management programme, and enable management to examine demand-side management programmes as they evolve.

7.8.3.2 Targeted end-use sectors/building types

Demand-side management programmes primarily target three end-use sectors: residential, commercial, and industrial. Each of these broad categories has multiple subsectors. In some cases, the programme will be designed for one or more broad sectors; in others, it may be tailored to a specific subsector. For example, the residential sector is divided into several subsectors, such as single-family, multi-family, mobile, and low-income housing.

7.8.3.3 Targeted end-use technologies/programme types

Some programmes are comprehensive and span multiple end-use technologies. Other programmes focus on specific end-use equipment, such as lighting, air conditioners, and dishwashers. Others focus on load control measures, such as demand response programmes, where customers temporarily reduce loads in response to peak demand events, or programmes that permanently shift loads to off-peak hours (e.g. via thermal energy storage).

7.8.3.4 Program implementers

Utility companies frequently implement demand-side management programmes. Other potential implementers include government organisations, non-profit groups, private parties, or a consortium of entities. Utilities and governments, in particular, have a vested interest in influencing customer demand, treating it as a choice rather than a fate, in order to provide better service at a lower cost while increasing profits and lowering business risks. Energy planners can use a variety of market push and pull strategies to influence consumer adoption and reduce barriers.

7.8.3.5 Implementation methods

One of the most important aspects of characterising demand-side alternatives is choosing the appropriate market implementation methods. Planners and policymakers can use a variety of methods to influence customer adoption and acceptance of demand-side management programmes.

The methods can be broadly classified into six categories.

1. **Customer education**: Many energy suppliers and governments have relied on some form of customer education to raise overall awareness of programmes. Websites, brochures, bill inserts, information packets, clearinghouses, educational curricula, and direct mailings are all frequently used. Customer education is the most fundamental market implementation method available, and it should be used in conjunction with one or more other market implementation methods to maximise effectiveness.
2. **Direct customer contact**: Direct customer contact techniques are those that involve face-to-face communication between a customer and an energy supplier or government representative in order to increase programme acceptance. Energy suppliers have long employed marketing and customer service representatives to advise on appliance selection and operation, heating/cooling system sizing, lighting design, and even home economics. Direct customer contact can be achieved through energy audits, specific programme services (for example, equipment servicing), storefronts that display information and devices, workshops, exhibits, on-site inspections, and so on. One significant advantage of these methods is that they enable the implementer to solicit consumer feedback, allowing them to identify and address major customer concerns. They also allow for more personalised marketing and can help communicate interest in and concern about energy cost control.
3. **Trade ally cooperation**: Trade ally cooperation and support can be critical to the success of many demand-side management initiatives. A trade ally is defined as any organisation that can influence transactions between a supplier and its customers, or between implementers and end users. Home builders and contractors, local chapters of professional societies, technology/product trade groups, trade associations, and associations representing wholesalers and retailers of appliances and energy-consuming devices are among the most important trade allies. Depending on the type of trade ally organisation, a variety of services are provided, such as standard and procedure development, technology transfer, training, certification, marketing/sales, installation, maintenance, and repair.

4. **Advertising and promotion**: Energy suppliers and government energy entities have employed a variety of advertising and promotion strategies. Advertising uses a variety of media to convey a message to customers in order to educate or persuade them. Demand-side management programmes can use advertising media such as the Internet, radio, television, magazines, newspapers, outdoor advertising, and point-of-purchase advertising. Press releases, personal selling, displays, demonstrations, coupons, and contests/awards are typical examples of promotional activities that support advertising.
5. **Alternative pricing**: Pricing as a market-influencing factor serves three purposes: (a) it informs producers and consumers about the cost or value of the products and services being provided; (b) it provides incentives to use the most efficient production and consumption methods; and (c) it determines who can afford how much of a product. These three functions are closely related. Alternative pricing, through innovative schemes, can be a valuable implementation technique for utilities that promote demand-side options. One significant advantage of alternative pricing programmes over other implementation methods is that the supplier incurs little or no cash outlay. The customer receives a financial incentive over a period of years, allowing the implementer to provide incentives while receiving benefits.
6. **Direct incentives**: Direct incentives are used to increase the short-term market penetration of a cost control/customer option by lowering the net cash outlay for equipment purchase or shortening the payback period (i.e. increasing the rate of return) to make the investment more appealing. Incentives also reduce customer resistance to options with no proven performance histories or that require significant changes to the building or the customer's lifestyle. Direct incentives include cash grants, rebates, buyback programmes, bill credits, and low- or no-interest loans. An additional type of direct incentive is the provision of free or heavily subsidised equipment installation or maintenance in exchange for participation. Such arrangements may cost the supplier more than the direct benefits of the energy or demand impact, but they can speed up customer recruitment and allow for the collection of useful empirical performance data.

The selection of the individual market implementation method or mix of methods depends on a number of factors, including the following:

1. Previous experience with similar programmes.
2. Existing market penetration.
3. The openness of policymakers and regulatory authorities.
4. Estimated programme benefits and costs for suppliers and customers.
5. Stage of buyer readiness.
6. Barriers to implementation.

Some of the most innovative demand-side marketing programmes began as pilot programmes to assess consumer acceptance and programme design before being implemented on a larger scale. The market implementation methods aim to influence the marketplace and change customer behaviour. The key question for planners and policymakers is how to select the market implementation method(s) that will result in the desired customer acceptance and response (Table 7.4). Customer acceptance refers to their willingness to participate in a market implementation programme, their decision to use the desired fuel/appliance choice and efficiency, and the behaviour change that the supplier or state encourages. Customer response refers to the actual load-shape change caused by customer action, as well as the characteristics of the devices and systems used. Customer acceptance and responses are influenced by demographics, income, knowledge, and awareness of available technologies and programmes, as well as decision criteria such as cash flow, perceived benefits and costs, attitudes, and motivations.

Table 7.4 Examples of market implementation methods.

Market implementation method	**Illustrative objective**	**Examples**
Customer education	Increase customer awareness of programs Increase perceived value of energy services	Websites Bill inserts Brochures Information packets Displays Clearinghouses Direct mailings
Direct customer contact	Through face to face communication, encourage greater customer acceptance and response to programs	Energy audits Direct installation Store fronts Workshops/energy clinics Exhibits/displays Inspection services

Table 7.4 *Continued.*

Market implementation method	Illustrative objective	Examples
Trade ally cooperation Trade ally cooperation(i.e., architects, engineers, appliance dealers, heating/cooling contractors)	Increase capability in marketing and implementing programs Obtain support and technical advice on customer adoption of demand-side technologies	Cooperative advertising and marketing Training Certification Selected product sales/service
Advertising and promotion	Increase public awareness of new programs Influence customer response	Mass media (Internet, radio, TV, and newspaper) Point-of-purchase advertising
Alternative pricing	Provide customers with pricing signals that reflect real economic costs and encourage the desired market response	Demand rates Time of use rates Real-time pricing Critical peak pricing Off-peak rates Seasonal rates Inverted rates Variable levels of service Promotional rates Conservation rates
Direct incentives	Reduce up-front purchase price and risk of demand-side technologies to the customer Increase short-term market penetration Provide incentives to employees to promoteside management programs	Low –or no-interest loan Cash grants Subsidised installation/modification Rebates Buyback programmes Rewards to employees for successful marketing of demand-side management programs

7.8.3.6 Characteristics of successful programs

7.8.3.6.1 Key elements of programme design

- Maintain simplicity in programme design.
- Design an incentive structure that meets the needs of the customer.
- Maintain a flexible programme design to accommodate measures.
- Maintain flexibility in programme design to meet market needs in specific industries.
- Develop effective performance tracking mechanisms.

Table 7.5 Synthesis of key challenges and success factors for selected energy efficiency and load management programs. (Source: Wikler.G. et al., Best practices in energy efficiency and load management programs, 1016383, EPRI, Palo Alto, CA, 2008.)
C&I = Commercial and industrial.

Program type	Key challenges	Key success factors
Residential conservation	Technology developments Easy access to pay stations	Enhancing customer knowledge and control over their consumption
Residential lighting	Low customer motivations, especially if electricity prices are low Utility resource allocation towards marketing and outreach efforts Overcome customer experience or perception of poor quality products CFL disposal issues	Aggressive customer awareness and outreach efforts Strong publicity campaigns Leverage ENERGY STAR brand Build and maintain relationships with manufacturers and retailers
Residential load management	Obtain customer trust to install equipment Address customer complaints unrelated to the program Low level of customer motivation if electricity prices are low	Simple program design Widespread customer awareness and outreach efforts Establish system reliability Maintain continuity in customer participation Build strong customer relationships Provide customer choice and control on electricity usage
Residential new construction	Generate interest in program by all parties Maintain consistency with builders and energy rates	ENERGY STAR name recognition Build and maintain relationships with builders and rates
Low income	Build customer trust and confidence Rapid turnaround in contractor workforce responsible for delivery	Deliver non-energy benefits such as comfort and safety Engender participant trust through program stability and continuity Successful education and outreach efforts Partner with community-based organisations for effective delivery

Table 7.5 *Continued.*

Program type	Key challenges	Key success factors
C & I energy efficiency	Incorporate flexible incentives to fit different customer requirements, while keeping program design simple	Simplicity in program design Strong financial incentives Keep customer's financial bottom-line in mind Strong customer relationships/maintaining close contact with customers directly and through contractors
C&I new construction	Generate interest from designers, builders, and owners Shortage of qualified staff	Build broad awareness Start with a pilot program and expand Staff development and training Flexible approach Full energy simulation capability
C&I retrofit	Need to conform to new/increased codes and standards Interactions with regulators	Flexibility in accommodating different measures Stakeholder collaborative process in designing programs Highly skilled utility staff Contractor-customer relationships
C&I niche	Accommodate industry requirements Obtain customer interest for participation Differences in using the new technology	Target niche, high-growth industries Dynamic program design to fit market requirements Form partnerships and collaborations with related groups Integrate energy efficiency into customers' business strategies Achieve more than just the energy benefits
Small business	Develop web-based delivery infrastructure Bring vendors up to speed	Make program participation as simple as possible Build a strong network of vendors

- Include customer choice and control features in programme design.
- Ensure resource allocation is commensurate with the programme tasks.
- Obtain stakeholder support right during the design stage.
- Maintain high-quality products.
- Establish programme branding.
- Implement programme improvements over time.

7.8.3.6.2 Key elements of programme delivery

- Encourage trade ally relationships and partnerships.
- Conduct contractor capacity building efforts.
- Establish networks and alliances with other relevant parties and organisations.
- Launch programme publicity campaigns.
- Develop strong customer education and outreach efforts.
- Strengthen utility-customer relationships.
- Coordinate with different utilities and programme administrators.
- Develop customer-contractor relationships.
- Maintain strong in-house capability.
- Integrate energy efficiency into customers' business strategies.
- Deliver non-energy benefits.
- Maintain consistency over time. Enhance programme delivery through collaboration.

7.8.4 Conclusion

Since the 1970s, various economic, political, social, technological, and resource supply factors have impacted the energy industry and its future outlook.

Many utilities are dealing with massive capital requirements for new plants, significant fluctuations in demand and energy growth rates, declining financial performance, and political, regulatory, and consumer concerns about rising prices and the environment. While demand-side management is not a panacea for these problems, it does provide numerous non-energy benefits in addition to the more obvious energy-related benefits. These demand-side alternatives should be considered by utilities, energy suppliers, energy-service providers, and government entities alike.

Implementing demand-side measures benefits the organisation by influencing load characteristics, delaying the need for new energy resources, improving resource value, and providing customers with benefits such as lower energy bills and improved performance from new technologies. Furthermore, society as a whole gains economic, environmental, and national security benefits. Demand-side management programmes, for example, can postpone the need for new power plants, saving money and reducing emissions associated with fossil-fuelled electricity generation. Demand-side management programmes also tend to generate more jobs and expenditures in the regions where they are implemented, thereby strengthening local economies. Furthermore, demand-side management programmes can help a country reduce its reliance on foreign oil imports, which improves national security. Demand-side management alternatives will continue to play an important role in resource planning in many countries, and they will be critical to achieving a sustainable energy future.

7.9 Thermal Energy Storage

This section will cover the basics of thermal energy storage. Deep discussion and details is beyond the scope of this section.

7.9.1 Introduction

The majority of energy management technology has focused on improving electricity consumption rather than timing demand. Some examples of consumption management devices include variable frequency drives, energy-efficient lighting, electronic ballasts, and energy-efficient motors. These techniques frequently have only a minor impact on the facility's demand (when compared to mechanical cooling equipment), which typically accounts for a significant portion of the facility's total annual electric bill. Demand charge management is primarily concerned with a generator's ability to supply power when needed, rather than energy conservation. The timing of consumption is the foundation of demand management and the focal point of thermal energy storage (TES).

Utilities frequently charge more for energy and demand during specific periods, such as on-peak rates and ratchet clauses. The process of managing the generation capacity that a specific utility has 'online' involves the utilisation of those generating units that produce power most efficiently first because these units would have the lowest avoided costs (ultimately the actual cost of energy). When loads approach the utility's connected generation capacity, additional generating units must be brought online. Each additional unit incurs an incrementally higher avoided cost because these 'peaking units' are less efficient and used less frequently. This has prompted many businesses to implement some form of demand management.

Thermal energy storage (TES) refers to the process of producing and storing energy in the form of heat or cold for use during peak periods.

For the profile shown in Figure 7.3, a cooling storage system could be used to reduce or eliminate the need to run the chillers during the peak rate period. Running the chillers during off-peak hours and storing this capacity for use during peak hours can result in energy cost savings. If this type of system is used during new construction or equipment replacement, smaller capacity chillers can be installed because the chiller can spread the production of the total load over the entire day rather than being sized for peak loads.

Thermal energy storage has been used for centuries, but it has only recently become popular among large electrical users for cost management. The process entails storing Btu (or a lack thereof) for use when a heat source or a heat sink is required. TES has expanded its applications by allowing

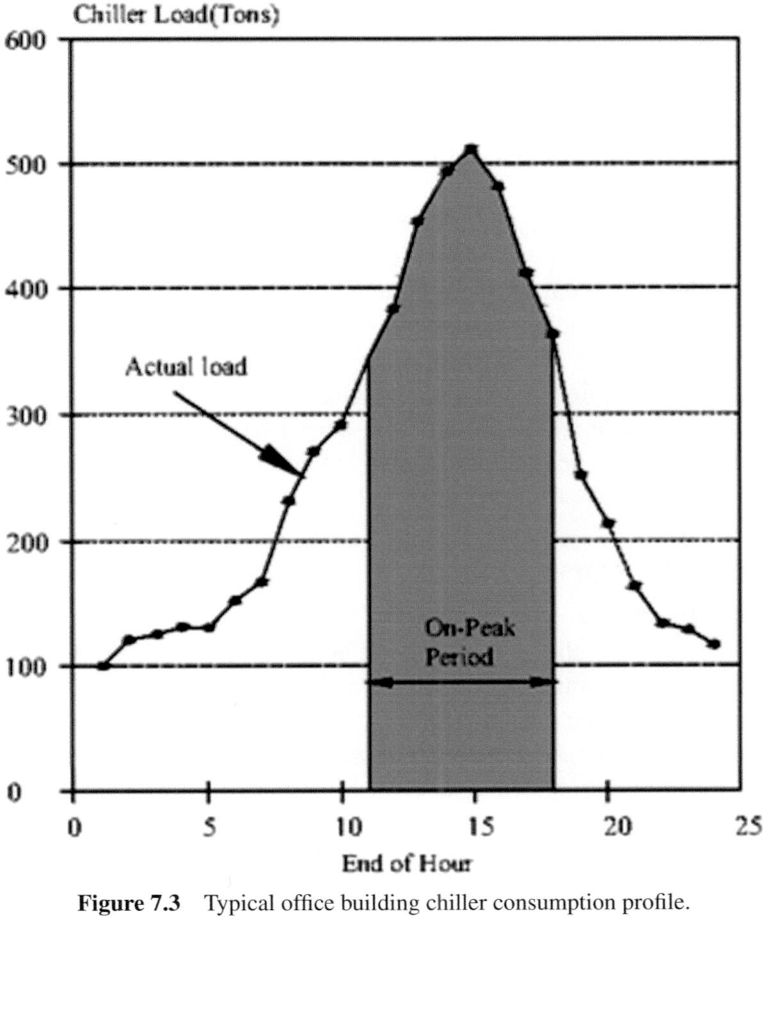

Figure 7.3 Typical office building chiller consumption profile.

for the use of low-cost energy sources (e.g. waste heat or time-based rate structure) to supplement more expensive energy sources later.

The discussion in this section will focus on the storage of cooling capacity; heat storage will not be addressed.

The chiller load and efficiency frequently follow the chiller consumption profile, with the chiller operating at a high load, i.e. high efficiency, for only a portion of the day. This is because the HVAC system must produce cooling when it is required, as well as handle instantaneous peak loads. Smaller chiller systems designed to handle base and peak loads during off-peak hours allow chillers to operate at higher average loads, resulting in higher efficiencies.

Thermal energy storage can also help to balance a cooling system's daily load. Conventional air conditioning systems require a chiller large enough to meet peak cooling demand as it occurs. This requires the cooling system to operate in load-following mode, varying its output in response to changes in cooling requirements. Systems that operate in a single or two shifts, as well as those that are heavily influenced by the weather, can benefit from TES' smoothing properties.

A school, for example, that adds a new wing could use its existing refrigeration system in the evening to generate cooling capacity that can then be stored for use during the day. Although the addition would require additional piping and pumping capacity, new chiller capacity may not be necessary. A new construction project with a similar single-shift cooling demand profile could use a smaller chiller in conjunction with storage to better balance the chiller operation. Moving load from the on-peak rate period to the off-peak period can help balance demand while also lowering residual ratcheted peak charges. Thermal energy storage is one method for accomplishing this.

7.9.2 Storage systems

Full storage systems

These ones turn off the chiller during peak hours and run entirely from the storage system. Full storage systems have a higher initial cost because the chiller is turned off during peak hours, and the cooling load must be met by a larger chiller running fewer hours, with an additional storage system storing the excess. Full storage systems save more than partial systems because the chillers are completely turned off during on-peak periods. Full storage systems are frequently used in retrofit projects because a large chiller system may already be in place.

Partial storage systems

These are intended to have the chiller run during on-peak hours, supplementing the storage system. A partial storage system offers significant savings with a lower initial cost and size requirements. New construction projects frequently use a partial storage system to reduce the size of both the chiller and the storage system. Partial load systems have the advantage of improving the performance of a system capable of handling the cumulative cooling load but not the building's instantaneous peak demands. In such a system, the chiller could operate continuously at a near-optimal load throughout the day, with excess cooling tonnage stored for use during peak periods. A system that already has two chillers is one option for using partial storage. The daily cooling load could be met by running both chillers during off-peak hours, storing any excess cooling capacity, and running only one chiller during peak hours to supplement storage system discharge. This also has the significant advantage of providing a reserve chiller during peak load periods. The partial load system may approach full load system characteristics early and late in the cooling season. As cooling loads and peaks begin to decline, the storage system will be able to handle more of the on-peak demand, and the on-peak chiller may eventually be turned off. A system like this can be designed to run the chillers at full capacity, increasing system efficiency.

Storage systems offer a variety of operational benefits to mechanical systems. Even a partially charged storage system could provide some capacity if the primary system fails or the utility goes down during an outage or a curtailment procedure. This 'redundancy' could be provided with minimal power, such as a small capacity emergency generator, because we could simply start chilled water pumps instead of a backup chiller to circulate water through a critical hospital system or data centre. With the right relationships in place, the utility could use a large storage system as a virtual generator during peak power periods.

7.9.3 Storage mediums

Several methods are currently in use to store cold in thermal energy storage systems. These include water, ice, and phase-change materials. The water systems simply store chilled water for use during peak periods. Ice systems generate ice that can be used to cool the chilling water by leveraging the high latent heat of fusion. Phase change materials are those with properties (such as melting points) that make them suitable for thermal energy storage.

7.9.3.1 Chilled water storage

Chilled water storage simply means storing chilled water generated during off-peak periods in a large tank or series of tanks. These tanks are the most common type of thermal storage. One reason for this popularity is the ease with which these water tanks can be integrated into the existing HVAC system. The chillers are not required to produce chilled water any colder than what is currently used in the system, so system efficiency is not compromised. The chiller system draws warm water from one end of the system and replaces it with chilled water at the other. During the off-peak charge cycle, the temperature of the water in the storage tank drops until it approaches or reaches the chiller system's output temperature. This chilled water is then withdrawn during the on-peak discharge cycle to supplement or replace the chiller(s) output.

Facilities with a system size constraint, such as a lack of space, frequently install a series of small insulated tanks connected in series. Other facilities have installed a single, high-volume tank, either above or below ground. The material and shape of these tanks vary greatly between installations. These large tanks are frequently designed in a way that is strikingly similar to municipal water storage tanks. The primary performance factors in the design of these tank systems, whether large or multiple, are location and insulation.

The advantages of using water as the thermal storage medium are:

1. Retrofitting the storage system to the existing HVAC system is simple.
2. Water systems use normal evaporator temperatures.
3. Water tanks have high thermal storage efficiency when properly designed.
4. Full thermal stratification maintains the chilled water temperature differential, thereby ensuring chiller loading and efficiency.
5. Water systems have lower auxiliary energy consumption than both ice and phase change materials because the water flows freely through the storage system.

7.9.3.2 Ice storage

Ice storage uses water's high latent heat of fusion to store cooling energy. One pound of ice contains 144 Btu of cooling energy, whereas chilled water contains only 1 Btu per pound. When ice systems are used instead of water, the required storage volume is reduced by about 75%. During off-peak periods, ice storage systems combine with the chiller system to form ice, which is then used to generate chilled water during peak periods.

The advantages of using ice as the thermal storage medium are:

1. Retrofitting the storage system with the existing HVAC chilled water system is possible.
2. Ice systems occupy less space than water systems.
3. Ice systems have more storage capacity but lower refrigeration efficiency than water systems.
4. Ice systems are available in packaged units due to their small size requirements.

7.9.3.3 Phase change materials

The use of phase change materials allows for the capture of latent heat from fusion while maintaining the evaporating temperatures of existing chiller systems. Materials with higher melting points than water have been successfully used in thermal energy storage systems. Several of these materials are classified as 'eutectic salts', which are salt hydrates made up of inorganic salts and water.

The advantages of using eutectic salts as the thermal storage medium are that they:

1. can utilise the existing chiller system for generating storage due to evaporator temperature similarity;
2. require less space than that required by the water systems;
3. have higher storage and equivalent refrigeration efficiencies to those of water; and
4. do not suffer the efficiency penalties of ice systems.

7.9.4 System capacity

The performance of thermal storage systems is dependent on proper design. If it is too small or too large, the overall system performance suffers. Full discussion of this is beyond the scope of this lecture.

7.9.5 Conclusion

Thermal energy storage will play a significant role in the future of demand-side management programmes for both private companies and utilities. To reduce utility costs, an organisation that wishes to implement a system-wide energy management strategy must be able to track, predict, and control its load profile. This management strategy will become even more important as electricity costs fluctuate in a deregulated market. Real-time pricing and multi-facility contracts will increase the savings potential of demand management, of which thermal energy storage should be an important tool.

The success of the thermal storage system and the HVAC system as a whole depends on many factors.

- The chiller load profile, the utility rate schedules and incentive programmes.
- The state of the existing chiller system.
- The space available for the different systems.
- The appropriate storage medium must be chosen, as well as the system's proper design and integration into the existing system.

Thermal storage is a very appealing way for a company to cut electricity costs and improve system management. Storage can be used in new installation projects to reduce both the initial costs of the chiller system and the operating costs. Storage systems will become more affordable in the future as mass production increases, technology advances, and more businesses transition to storage.

References

[1] Wikler.G. et al., Best practices in energy efficiency and load management programs, 1016383, EPRI, Palo Alto, CA, 2008.

[2] IEA (2023), Energy Efficiency 2023, IEA, Paris https://www.iea.org/reports/energy-efficiency-2023, Licence: CC BY 4.0. (Accessed: 08 May 2024)

[3] IEA (2023), World Energy Outlook 2023, IEA, Paris https://www.iea.org/reports/world-energy-outlook-2023, Licence: CC BY 4.0 (report); CC BY NC SA 4.0 (Annex A). (Accessed: 08 May 2024)

[4] Barney L. Cape hart, Wayne C. Turner (2016), Guide to Energy Management, Fairmont Press, ISBN: 1420084895

[5] Clive Beggs (2002), Energy: Management, Supply and Conservation, Butterworth Heinemann, ISBN: 0750650966

[6] Wayne C. Turner & Steve Doty (2012), Energy Management Handbook, 7th Ed., Fairmont Press, ISBN: 142008870X

[7] *Case Studies by System. Energy.gov*. Available at: https://www.energy.gov/eere/iedo/case-studies-system.

[8] *Case studies by System | Department of Energy*. Available at: https://www.energy.gov/eere/iedo/case-studies-system (Accessed: 08 May 2024).

[9] *Smart energy management for industrials, Deloitte Insights*. Available at: https://www2.deloitte.com/us/en/insights/industry/power-and-utilities utilities/smart-energy-management.html. (Accessed: 08 May 2024)

[10] Waters, J.R. (2008) Energy Conservation in Buildings. John Wiley & Sons.

[11] Kreith, F. and D. Yogi Goswami (2016) Energy Management and Conservation Handbook, Second Edition. CRC Press.

[12] John Littler, Randall Thomas – (1984) Design with Energy_ The Conservation and Use of Energy in Buildings -Cambridge University Press.

[13] Panke, R. (2001). Energy Management Systems & Direct Digital Control (1st ed.). River Publishers. https://doi.org/10.1201/9781003150992.

[14] *Saving Energy in Industrial Companies: Case Studies of Energy Efficiency Programs in Large U.S. Industrial Corporations and the Role of Ratepayer-Funded Support | Department of Energy*. Available at: https://www.energy.gov/sites/default/files/2021-07/saving_energy_industrials.pdf (Accessed: 08 May 2024).

[15] A. Goldberg, R.P. Taylor, and B. Hedman, Industrial Energy Efficiency: Designing Effective State Programs for the Industrial Sector," State and Local Energy Efficiency Action Network, Industrial Energy Efficiency and Combined Heat and Power Working Group, 2014, (Accessed: 08 May 2024) http://energy.gov/sites/prod/files/2014/03/f13/industrial_energy_efficiency.pdf. (Accessed: 08 May 2024)

8

Transportation and Energy Conservation

Chapter Preview

Efficient transportation is critical to the success of modern economies. However, oil reserves appear to be dwindling, raising concerns about the sustainability of many modes of transportation. In this chapter we start by a brief history of transport and its relation to economy. We delve into energy consumption and transport, sustainable transport options, urban planning and sustainable transportation, fuel efficiency, and alternative fuels.

8.1 Transport and the Economy

Transportation is important. Imagine life without internal combustion or jet engines. Getting to work can be difficult, especially if your home is a long distance from your workplace. Similarly, visiting family and friends would be significantly more difficult. This, however, would be the tip of the iceberg. How would the countless food producers get their products on the shelves of your neighbourhood supermarket? Transport is a critical issue for governments around the world; without efficient transportation systems, economies struggle to grow because labour and goods cannot be easily moved around. Efficient transportation is therefore critical for all societies because it influences both economic development and population welfare. Efficient transportation systems provide numerous economic and social benefits, whereas inefficient systems incur economic costs in the form of missed or reduced opportunities.

Improved transport systems can benefit economies in two ways.

- For starters, they provide people with access to places where they can generate wealth and consume goods and services, such as educational and healthcare facilities. This ultimately leads to a healthier, more educated society and allows for the development of larger markets, saving time and money.

- Second, transportation allows businesses to easily access raw materials and parts, thereby lowering production costs. It also allows them to deliver their products to customers more quickly. In doing so, it serves as an intermediate input for production. Furthermore, transportation serves an important social function, allowing people to network and socialise, thereby promoting the growth of leisure services and facilities.

Many economists believe that transportation promotes mobility, which is a reliable indicator of economic development. Mobility addresses one of the most fundamental aspects of economic activity: the need to transport people, goods, and information from one location to another. As a result, economies with greater mobility have more opportunities for development than those with limited mobility.

Reduced mobility stifles economic development, whereas increased mobility serves as a catalyst for growth.

8.2 Brief History of Transport

Overland travel has been a difficult and dangerous business throughout human history. Mountains, forests, ravines, and other geographical features made land transportation a slow and difficult experience. In comparison, travelling by boat along rivers and seas was far easier. As a result, early human settlements grew up along rivers and coastlines. Over time, this resulted in the establishment of large seaports, allowing nations to conduct international trade. These seaports enabled sailors to travel the world in large vessels, allowing European maritime nations to prosper from international trade through their colonial empires. Water transport was also essential in the early stages of the Industrial Revolution. In the eighteenth and early nineteenth centuries, the development of canal systems in Western Europe made it possible for businesses to transport heavy goods over long distances relatively easily. They enabled manufacturing centres to thrive in inland areas for the first time.

Railways first appeared in the middle of the nineteenth century. These were the first integrated inland transportation system, allowing for mass transport of people and goods, surpassing the flexibility of previous canals. Railway expansion fuelled the growth of industrial centres in North America and Europe. The twentieth century saw the rise of road transportation and automobile manufacturing. While this improved individual transportation for the masses, it was not until after WWII that automobile ownership became widespread. As car ownership increased, highway infrastructure improved, resulting in significantly shorter travel times.

The later part of the twentieth century saw the rise of global air travel. This allowed the masses for the first time to travel abroad, facilitating growth in international trade. The effect of this was to break down international boundaries and make the world smaller, which was reinforced by the explosion in telecommunications in the late twentieth century.

The history of transportation mirrors that of global economic development. In a few hundred years, the world has evolved from a collection of isolated societies to an interconnected network dominated by global corporations. International trade is central to the global economy, with countries such as China emerging as global manufacturing powerhouses. Without an efficient international transportation system, the global economy would struggle to function. Indeed, the world's economies are now so reliant on international trade that it is not surprising that there is a strong vested interest in ensuring that international transport continues to thrive.

8.3 Passenger Transport

The amount of energy used for transportation depends on the distance travelled by passengers and goods, as well as the preferred mode of transportation. Transport systems follow supply and demand laws similar to other industries, but are further complicated by network effects and modes of transportation. Comfort and convenience are key considerations.

8.4 Energy Consumption and Transport

Internal combustion engines power the majority of vehicles that travel over land, as well as many ships. Such engines use oil as fuel and are notoriously inefficient. A typical petrol engine has a thermal efficiency of approximately 26% before accounting for mechanical inefficiencies. This means that most standard engines have an overall efficiency of around 20%. In other words, 80% of the fuel's energy is lost to the atmosphere as heat. In comparison, diesel engines with a higher compression ratio than petrol engines have efficiencies of around 45%, making them a far more energy-efficient option. Automobile manufacturers have made significant efforts to improve the efficiency of petrol engines, as internal combustion engines are inefficient. However, despite significant improvements in engine efficiency, the average fuel efficiency of gasoline-powered vehicles has remained relatively stable, owing to the fact that more people now own large luxury vehicles.

Car and train manufacturers prioritise mechanical efficiency, but the way vehicles are used heavily influences overall efficiency, which is entirely user-driven. Comparing energy performance between vehicle types is challenging due to the impact of user-related variables. While mechanical efficiency is important, what matters most is the energy consumed per passenger kilometre, which is entirely dependent on how full cars, buses, and trains are.

If a large intercity train carries only a few passengers, the fuel consumption per passenger-kilometre will be significantly higher than if it is full. Comparing energy efficiency between modes of transportation can be challenging due to unrealistic assumptions about passenger load and engine size.

8.5 Sustainable Transportation

Sustainable transportation is critical in the efforts to conserve energy and protect the environment. Sustainable transportation alternatives provide a way to reduce emissions and rely less on non-renewable energy sources. This section investigates various sustainable transportation modes, such as electric vehicles, public transit, cycling, and walking, and emphasises the role of urban planning in promoting these environmentally friendly options.

8.5.1 Electric vehicles: Leading the charge in green transportation

Electric vehicles (EVs) are at the forefront of the transition to sustainable transportation.

- Environmental advantages: Compared to conventional vehicles, EVs emit significantly less greenhouse gas emissions, particularly when charged with renewable energy.
  - Zero emissions: Electric vehicles have no tailpipe emissions, which reduces air pollution and greenhouse gas emissions.
  - Energy efficiency: Electric vehicles (EVs) consume less energy than internal combustion engines (ICEs).
  - Renewable energy compatibility: EVs contribute to a greener energy grid when charged with renewable energy sources (such as solar or wind).

- Technological advancements: Continuous improvements in battery technology, as well as the expansion of charging infrastructure, are making electric vehicles more accessible and convenient.
- Policy support: Government incentives and regulations are also driving EV adoption, highlighting their importance in sustainable transportation.
- **Challenges and considerations**:
    - Infrastructure: Expanding charging infrastructure is critical to widespread EV adoption.
    - Battery materials: The proper disposal and recycling of lithium-ion batteries is critical.
    - Range anxiety: Improving battery technology to extend driving range is a continuous challenge.

8.5.2 Public transit: Efficient and eco-friendly

Public transportation systems, such as buses, trains, and trams, are critical components of sustainable urban mobility.

- Carbon footprint reduction: By replacing private car trips with efficient public transportation, individuals can significantly reduce their carbon emissions.
- Congestion reduction: Well-designed public transportation networks reduce traffic congestion, resulting in lower total urban emissions.
- Successful examples: Cities such as Amsterdam, Singapore, and Tokyo demonstrate the importance of effective public transportation in promoting sustainable mobility.
- **Environmental benefits**:
    - **Energy efficiency**: Buses and trains are more energy-efficient when transporting a large number of passengers.
    - Reduced congestion: Public transportation reduces the number of individual cars on the road, thereby alleviating traffic congestion.
    - Lower emissions: Mass transit emits fewer pollutants per passenger than private vehicles.
- **Urban planning**: Effective urban planning involves:
    - Integrated systems: Connecting public transportation routes to residential and commercial areas.
    - Accessibility: Providing easy access to transit stops.
    - Affordability: Making public transportation affordable to all socioeconomic groups.

8.5.3 Cycling and walking: Zero-emission transportation

Cycling and walking are the most energy-efficient and eco-friendly transportation methods:

- Health and environmental benefits: These zero-emission modes of transportation reduce air pollution while also improving physical health and well-being.
- Infrastructure support: Creating dedicated bike lanes, pedestrian zones, and greenways is critical for promoting sustainable transportation options.
- **Environmental benefits**:
    - Zero emissions: Cycling and walking have no emissions.
    - Reduced traffic congestion: Fewer cars on the road mean less congestion.
    - Healthy cities: Promoting active transportation benefits public health.
- **Urban design**: Creating bike lanes, pedestrian pathways, and safe crossings improves long-term mobility.

8.6 Urban Planning and Sustainable Transportation

The design and planning of urban spaces significantly influence transportation choices:

- Walkable and bike-friendly cities: Integrating cycling and walking paths into city infrastructure encourages their use while also improving urban liveability.
- Public transit integration is a key component of effective urban planning.
- The 'Complete Streets' concept involves designing streets to accommodate all users, including pedestrians, cyclists, motorists, and transit riders, resulting in a more sustainable and inclusive transportation ecosystem.

Role of urban planning

Urban planning plays a critical role in promoting sustainable transportation:

- Mixed-use development: Creating neighbourhoods with a mix of residential, commercial, and recreational spaces promotes walking and reduces the need for lengthy commutes.

- Transit-oriented development (TOD): Centring development around transit hubs promotes public transportation use.
- Zoning policies: Implementing zoning regulations that prioritise pedestrian-friendly areas while discouraging car-centric designs.

8.7 Fuel Efficiency and Alternative Fuels

Fuel efficiency and alternative fuels play critical roles in transportation and energy conservation. Increasing fuel efficiency in conventional vehicles and developing alternative fuel sources such as biofuels, hydrogen, and electricity are critical for lowering energy consumption and emissions.

8.7.1 Improving fuel efficiency in conventional vehicles

To improve fuel efficiency in conventional vehicles, several strategies are employed:

- Aerodynamics and lightweight materials: Advancements in vehicle design reduce air resistance, while lightweight materials improve fuel efficiency.
- Advanced engine technologies: Technologies such as variable valve timing and turbocharging improve engine efficiency.
- Hybrid technologies: Combining conventional engines with electric power significantly reduces fuel consumption, representing a practical step towards more sustainable transportation.

Fuel efficiency is critical for reducing energy consumption and vehicle emissions. Strategies to improve fuel efficiency in conventional vehicles:

- Aerodynamics: Streamlined vehicle designs reduce air resistance, which improves fuel efficiency. Smooth contours, underbody panels, and active grille shutters improve aerodynamics.
- Lightweight materials: Using lightweight materials (such as aluminium, carbon fibre, and high-strength steel) reduces overall vehicle weight. Lighter cars use less energy to accelerate and maintain speed.
- Engine efficiency: Internal combustion engines (ICEs) have improved significantly in terms of fuel efficiency. Direct fuel injection, variable valve timing, and cylinder deactivation are among the most significant advancements.
- Transmission optimisation: Modern transmissions (automatic, continuously variable, or dual-clutch) optimise gear shifts, allowing engines to operate efficiently under a variety of driving conditions.

- Tyre pressure maintenance: Properly inflated tyres reduce rolling resistance, which improves fuel economy. Regularly checking and maintaining tyre pressure is essential.
- Driving habits: Encouraging environmentally friendly driving practices, such as smooth acceleration, avoiding aggressive braking, and maintaining a constant speed, can have a significant impact on fuel efficiency.

8.7.2 Biofuels: A renewable energy source

Biofuels present a renewable alternative to fossil fuels:

- Common biofuels include ethanol, which is produced from crops such as corn and sugarcane, and biodiesel, which is derived from vegetable oils and animal fats.
- Environmental advantages: Biofuels can reduce greenhouse gas emissions, which contributes to a lower environmental impact.
- Production challenges: Land use, competition with food production, and the overall energy balance of biofuel production all present significant challenges.

8.7.3 Hydrogen fuel: The future of zero-emission vehicles

Hydrogen fuel cells offer a vision for zero-emission vehicles:

- Hydrogen fuel cells generate electricity for vehicles while emitting only water vapour.
- Environmental advantages: Hydrogen, as a clean energy source, has the potential to significantly reduce transportation emissions.
- Adoption challenges: Developing the necessary fuelling infrastructure and sustainable hydrogen production methods are currently impediments to broad adoption.

8.7.4 Electricity: Driving the transition to clean energy

Electric vehicles (EVs) are at the forefront of the transition to clean energy in transportation:

- Advancements in EV technology: Advances in battery technology, such as longer life and faster charging, are increasing the utility of EVs.
- Infrastructure development: Expanding charging infrastructure is critical for increasing EV adoption.

- Energy sourcing and grid capacity: The challenges include ensuring a clean energy source for electricity and managing increased demand on power grids.

8.7.5 Alternative fuels

1. **Biofuels**:
   - They are made from organic materials (like corn, sugarcane, or algae) and include ethanol (E85) and biodiesel.
   - Ethanol is typically blended with gasoline and used in flexible-fuel vehicles (FFVs).
   - Biodiesel made from vegetable oils or animal fats can be used instead of or in addition to diesel fuel.
2. **Hydrogen fuel cells**:
   - Hydrogen fuel cells produce electricity by combining hydrogen and oxygen while emitting only water vapour.
   - FCVs use hydrogen as a clean energy source with no tailpipe emissions.
3. **Electric vehicles (EVs)**:
   - EVs rely solely on electricity stored in batteries.
   - Battery technology advancements have resulted in longer driving ranges and shorter charging times.
   - Infrastructure development (charging stations) and battery recycling are two significant challenges.
4. **Natural gas**:
   - CNG and LNG are cleaner alternatives to gasoline and diesel.
   - Commonly used in buses, trucks, and some passenger vehicles.
5. **Synthetic fuels**:
   - Synthetic fuels (such as synthetic gasoline or diesel) are made from non-petroleum feedstocks.
   - They have the potential to reduce greenhouse gas emissions.

8.7.6 Technological advancements and challenges

1. **Advancements**:
   - **Advanced combustion engines**: Further research into more efficient ICEs.

- **Hybridisation**: Combining internal combustion engines and electric powertrains.
- **Autonomous driving**: Autonomous vehicles can save fuel by planning their routes more efficiently.
- **Materials science**: Developing lightweight and durable materials.
- **Smart grid integration**: EVs as grid resources for energy storage and load management.

2. **Challenges**:
   - **Infrastructure**: Building charging stations and hydrogen refuelling stations.
   - **Cost**: EVs and fuel cell vehicles are often costlier upfront.
   - **Range anxiety**: Addressing concerns about EV driving range.
   - **Policy and regulation**: Encouraging adoption through incentives and regulations.

8.7.7 Reducing energy consumption in transportation

The transportation sector has a significant impact on global energy consumption and emissions. Reducing its energy consumption is critical to achieving broader energy conservation objectives. Here are strategies and practices for managing transportation demand, technological innovations for efficient routing, and policies that promote energy-efficient practices, with an eye towards their potential impact on energy conservation.

8.7.8 Transportation demand management (TDM)

Transportation demand management aims to maximise the use of existing transportation infrastructure:

- TDM strategies include promoting public transportation, carpooling programmes, and flexible work arrangements to reduce peak traffic congestion.
- TDM aims to reduce overall vehicle miles travelled by encouraging modal shifts from single-occupancy vehicles to more efficient modes, resulting in lower energy consumption.
- Impact on energy conservation: Effective TDM can significantly reduce fuel consumption and emissions, thereby contributing to overall energy conservation efforts.

Technology for efficient routing and logistics

Technological advancements are revolutionising routing and logistics in transportation:

- **Advanced navigation tools**: GPS and AI-based systems make routing more efficient, reducing unnecessary travel and idle time.
- **Application in freight transport**: In freight transport, these technologies optimise delivery routes, resulting in significant fuel savings.

Policies encouraging energy-efficient practices

Policies play a crucial role in promoting energy efficiency in transportation:

- Regulatory measures include imposing fuel efficiency standards on vehicles and offering subsidies or incentives for purchasing electric vehicles.
- Governments are investing in green infrastructure, including electric vehicle charging stations and improved public transportation networks.
- Challenges and considerations: While these policies are beneficial, they must be carefully designed to ensure feasibility and equitable access to sustainable transportation alternatives.

8.7.9 Impact of strategies on energy conservation

TDM, technological innovations, and supportive policies all contribute to a significant reduction in transportation energy consumption. These strategies result in direct energy savings while also providing broader environmental and societal benefits, such as reduced air pollution and improved urban liveability.

Strategies for reducing energy consumption in the transportation sector

1. **Keep motorisation growth rates in check**:
    - The world is currently on track to increase the number of cars from 1.3 billion today to a staggering 2.2 billion by 2050. Despite the rise of electric vehicles, curbing motorisation remains crucial for reducing transportation-related oil use and carbon emissions.
    - Addressing this challenge requires individual behaviour changes and a shift away from decades-old motorisation dependence. Governments, businesses, and individuals must prioritise alternatives to

private car ownership, such as public transportation, cycling, and walking.

2. **Invest in public transport and active mobility**:
   - **Public transport**: Enhance public transportation systems by expanding capacity and improving accessibility. Doubling public transport capacity and aiming for 50% of trips via walking or cycling by 2030 can help limit global temperature rise to 1.5 °C, a critical threshold for mitigating climate change.
   - **Walking and cycling infrastructure**: Create safe and convenient pedestrian lanes, bike pathways, and car-free zones. These investments yield benefits for the climate, economy, and human health.
   - Despite these clear advantages, public transport and active mobility options remain underfunded compared to car infrastructure.
3. **Compact land use and car-free infrastructure**:
   - Cities can reduce transport-related fuel consumption by approximately 25% through smart land use planning. Compact urban development, combined with pedestrian-friendly infrastructure, reduces the need for long car trips.
   - Prioritising pedestrian lanes, bike paths, and car-free zones encourages energy-efficient modes of travel while minimising reliance on fossil fuels.
4. **Energy-saving tips for individuals**:
   - **Limit private vehicle use**: Whenever possible, opt for public transportation instead of driving your own car.
   - **Reduce car weight**: Remove unnecessary items from your car's trunk (like a spare tire or heavy luggage) to improve fuel efficiency.
   - **Regular vehicle maintenance**: Keep your car's fuel filters clean and ensure regular tune-ups.
   - **Avoid using the clutch pedal as a footrest**: This small change can save fuel.
   - **Restart your car**: If idling for more than a minute, it's more fuel-efficient to turn offoff and restart your vehicle.
   - **Maintain recommended tire pressure**: Properly inflated tires improve fuel economy.

5. **Global efforts to reduce emissions**:
   - To limit global warming to 1.5 °C, we must rapidly decrease transportation emissions. Achieving this goal involves a combination of strategies, including those mentioned above, to create a more sustainable and resilient planet.

References

[1] Wikler.G.et al., Best practices in energy efficiency and load management programs, 1016383, EPRI, Palo Alto, CA, 2008.
[2] IEA (2023), Energy Efficiency 2023, IEA, Paris https://www.iea.org/reports/energy-efficiency-2023, Licence: CC BY 4.0. (Accessed: 08 May 2024)
[3] IEA (2024), Global EV Outlook 2024, IEA, Paris https://www.iea.org/reports/global-ev-outlook-2024, Licence: CC BY 4.0. (Accessed: 08 May 2024)
[4] IEA (2023), World Energy Outlook 2023, IEA, Paris https://www.iea.org/reports/world-energy-outlook-2023, Licence: CC BY 4.0 (report); CC BY NC SA 4.0 (Annex A). (Accessed: 08 May 2024)
[5] El-Samra, S. and Adriazola-Steil, C. (2022) '5 Ways to Cut Oil and Gas Use Through Clean Transportation', *www.wri.org* [Preprint]. Available at: https://www.wri.org/insights/5-ways-cut-oil-and-gas-use-through-clean-transportation. (Accessed: 08 May 2024)
[6] *5 ways for the world to reduce emissions from global transport systems* (2023) *World Economic Forum*. Available at: https://www.weforum.org/agenda/2023/03/5-ways-the-world-can-reduce-emissions-from-global-transport-systems/. (Accessed: 08 May 2024)
[7] El-Samra, S. and Adriazola-Steil, C. (2022) '5 Ways to Cut Oil and Gas Use Through Clean Transportion', *www.wri.org* [Preprint]. Available at: https://www.wri.org/insights/5-ways-cut-oil-and-gas-use-through-clean-transportation. (Accessed: 08 May 2024)
[8] Strengthening International Cooperation for a Global Energy Transition | Climate-Diplomacy. Available at: https://climate-diplomacy.org/sites/default/files/2020-10/IASS_Policy_Brief_2019_StrengtheningInternationalCooperationforaGlobalEnergyTransition.pdf (Accessed: 08 May 2024).

[9] Inclusive and Sustainable Industrial Development Practical Guide for Implementing an Energy Management System, Available at: https://www.unido.org/sites/default/files/2017-11/IEE_EnMS_Practical_Guide.pdf.

[10] US EPA,OAR (2017) Quantifying the Multiple Benefits of Energy Efficiency and Renewable Energy: A Guide for State and Local Governments | US EPA, US EPA. Available at: https://www.epa.gov/statelocalenergy/quantifying-multiple-benefits-energy-efficiency-and-renewable-energy-guide-state. (Accessed: 08 May 2024)

9

Policy and Regulations

Chapter Preview

This chapter briefly reviews policy and regulations related to energy conservation and management. We look at government initiatives and incentives and international agreements on energy conservation.

9.1 Government Initiatives and Incentives

Government policies and incentives are critical in encouraging energy conservation and the use of renewable energy. Governments around the world are encouraging individuals and businesses to adopt energy-efficient practices and technologies through a variety of financial mechanisms, including tax credits, rebates, and grants. This section examines these incentives, emphasising their effectiveness and impact on promoting sustainable energy use.

Tax credits for energy efficiency

Tax credits are a powerful motivator for implementing energy-efficient solutions. These credits can help pay for some of the costs of installing renewable energy systems or upgrading residential and commercial properties to be more energy efficient. For example, homeowners may be eligible for tax credits for installing solar panels or upgrading to energy-efficient windows, whereas businesses may be eligible for credits for implementing green energy systems. These financial incentives not only lower initial investment costs, but also promote widespread adoption of energy-efficient technologies.

Rebates and grants for renewable energy projects

Rebates and grants are important in promoting renewable energy projects. Rebate programmes frequently cover a portion of the cost of installing renewable energy systems, such as solar or wind power installations. Furthermore, grants are available to fund larger-scale renewable energy projects or innovative research into new energy technologies. These grants can help start-ups and research institutions develop new sustainable energy solutions.

Incentives for energy-efficient appliances and systems

Many governments provide incentives to purchase energy-efficient appliances and systems. These could include discounts, rebates, or tax breaks on items like energy-efficient HVAC systems, lighting, and refrigerators. These programmes seek to lower the barrier to purchasing energy-efficient appliances by encouraging consumers and businesses to invest in technologies that reduce energy consumption and long-term operational costs.

Policy initiatives and regulatory support

Beyond individual incentives, broad policy initiatives and regulatory support are critical. This includes establishing renewable energy standards, requiring energy audits, and enforcing building and appliance efficiency regulations. Regulatory bodies play an important role in enforcing these policies, ensuring standards are met, and encouraging widespread energy conservation.

Challenges and future directions

While these initiatives have significant impact, they face challenges such as ensuring equitable access and adapting to rapidly evolving technologies. Future directions could include more targeted incentives and integrated policy approaches to improve energy efficiency and renewable energy use.

Government initiatives and incentives for energy conservation and renewable energy

1. **Federal programs**
    - **Department of energy office of energy efficiency and renewable energy (EERE)**:
        - **Renewable energy**:

  - The *Bioenergy Technologies Office* (formerly the Biomass and Bio refinery Systems R&D Program) focuses on advancing biomass utilisation.
  - *Regional Biomass Energy Grant Programs* support local projects related to biomass energy.
  - The *Geothermal Technologies Office (GTO)* promotes geothermal energy development.
  - *Hydrogen & Fuel Cell Technologies Office* encourages research and deployment of hydrogen and fuel cell technologies.
  - *Solar Energy Technologies Office (SETO)* drives solar energy innovation.
  - The *Water Power Technologies Office* (formerly Wind and Hydropower Technologies Program) explores hydropower and marine energy.
  - The *Wind Energy Technologies Office* (formerly Wind and Hydropower Technologies Program) advances wind energy research.
- **Energy efficiency**:
  - *Building Technologies Office* focuses on energy-efficient building designs and technologies.
  - The *Weatherization Assistance Program (WAP)* assists low-income households in improving energy efficiency.
  - The *Advanced Manufacturing Office (AMO)* (formerly the Industrial Technologies Program – ITP) supports energy-efficient industrial processes.

2. State and local programs

- **Rebates, tax credits, and savings programs**:
  - Consumers can find financial incentives for energy-efficient and renewable energy products. These include rebates, tax credits, and financing programs.
  - State and local governments often offer additional incentives tailored to their regions.

3. Global initiatives

- *Households across the world* benefit from government initiatives that provide rebates, credits, or discounts on renewable energy technologies. These efforts accelerate the transition to low-carbon energy.

9.2 International Agreements on Energy Conservation

International cooperation is essential for addressing the complex issues of global energy conservation and climate change. International agreements commit countries around the world to collective action in order to shape a sustainable future. These agreements play an important role in setting global standards and influencing national policies, ensuring a coordinated response to the pressing issue of climate change.

The Paris Agreement: A landmark in climate action

- **Paris Agreement**:
  - The *Paris Agreement*, adopted in 2015, represents a landmark global commitment to combat climate change. Its primary goal is to limit global warming to well below 2 °C above pre-industrial levels, with efforts to limit it to 1.5 °C.
  - **Nationally determined contributions (NDCs)**: Each signatory country submits its NDC, outlining its specific targets for reducing greenhouse gas emissions and adapting to climate impacts. These contributions are crucial for achieving the Paris Agreement's objectives.
  - **Renewable energy targets**: Many NDCs include quantified targets for renewable energy adoption. If all these targets were implemented, an additional 1041 gigawatts of renewables would be added by 2030, significantly reducing energy-related CO_2 emissions.
  - **IRENA's role**: The International Renewable Energy Agency (IRENA) supports countries in enhancing their NDCs through accelerated deployment of renewables. IRENA collaborates with 66 countries, including small islands and least developed nations, to scale up renewable energy initiatives.

The Paris Agreement stands as a landmark in international efforts to combat climate change. Key aspects include:

- Global warming limit: The agreement aims to keep global warming well below 2 °C above pre-industrial levels, with efforts to limit it to 1.5 °C.

- Nationally determined contributions (NDCs): Each participating country establishes its own targets for emissions reduction and renewable energy, based on its capacity and circumstances.
- Regular review and update: Every five years, countries must revisit and enhance their NDCs to ensure progressive ambitions in climate action. The Paris Agreement's adaptable yet ambitious framework encourages countries to implement more stringent energy-saving measures and accelerate the transition to renewable energy.

Other significant international agreements and protocols

Beyond the Paris Agreement, several other international accords contribute to global energy conservation efforts:

- **The Kyoto Protocol**: Although not as active today, the Kyoto Protocol (1997) set binding emission reduction targets for industrialised countries.
- **International Renewable Energy Agency (IRENA)**: An intergovernmental organisation supporting countries in their transition to sustainable energy. While varying in approach and scope, these initiatives collectively underscore the global commitment to reducing greenhouse gas emissions and fostering sustainable energy practices.
- **Montreal Protocol**: Focused on phasing out ozone-depleting substances, the Montreal Protocol (1987) indirectly contributes to energy conservation by promoting alternatives.
- **IEA agreements**: The International Energy Agency (IEA) facilitates cooperation among member countries to enhance energy security, efficiency, and sustainability.
- **Bilateral engagements**: Countries engage bilaterally to share lessons and promote sustainable energy transitions.

Challenges and opportunities in international collaboration

International collaboration in energy conservation is not without challenges:

- Diverse national interests: Balancing competing economic and political interests can be difficult.
- Implementation complexity: Translating global commitments into effective national policies requires significant effort. However, these collaborations provide opportunities for technology sharing, joint

research initiatives, and financial mechanisms to support energy transition in developing countries.

9.2.1 Impact on national energy policies and global climate response

International agreements have a significant impact on national energy policies. They drive legislative and regulatory changes that promote renewable energy, energy efficiency, and greenhouse gas emissions reductions. These agreements demonstrate a shared commitment to addressing climate change by encouraging innovation, investment in clean energy, and more sustainable practices.

Energy efficiency regulations

For businesses and industries, complying with energy efficiency regulations is both a legal requirement and an opportunity to embrace sustainable practices. These regulations, which range from local building codes to international standards, are intended to reduce energy consumption while promoting environmental responsibility. Understanding and adhering to these standards is critical for any organisation that values sustainability and operational efficiency.

Understanding energy efficiency standards and regulations

Energy efficiency regulations encompass a wide range of requirements:

- Building codes establish energy efficiency standards for building design, construction, and operation.
- Appliance and equipment standards: Specifications for the energy efficiency of appliances and industrial equipment.
- Industrial benchmarks are specific performance standards for different industries. Staying informed about these regulations, which can vary greatly by region and industry, is critical for ensuring compliance.

Benchmarking and assessment

Benchmarking involves comparing an organisation's energy usage against standards or industry averages. Key steps include:

- Energy audits: Conducting comprehensive energy audits to assess current energy consumption and identify opportunities for improvement.

- Performance metrics: Creating metrics to compare energy performance to regulatory benchmarks. This assessment is essential for developing strategies to meet or exceed energy efficiency standards.

Reporting requirements and documentation

Many energy efficiency regulations require regular reporting and documentation, including:

- Energy use data: Comprehensive records of energy consumption and efficiency measures.
- Compliance reports: Compliance reports are periodic submissions that demonstrate adherence to regulations. Maintaining accurate and up-to-date records is critical for demonstrating compliance and avoiding penalties.

Strategies for ensuring compliance

Ensuring compliance with energy efficiency regulations involves several strategies:

- Energy management systems: Systems that monitor and manage energy consumption.
- Training and awareness: Educating employees about energy-efficient practices and regulatory requirements.
- Regular policy review: Constantly reviewing and updating energy policies to ensure compliance with changing regulations and technological advancements.

Dealing with non-compliance and penalties

Non-compliance can result in legal penalties, fines, and reputational damage. Understanding the consequences of non-compliance and taking proactive steps to follow regulations is critical.

9.2.2 Navigating energy efficiency standards and regulations

1. **Benchmarking:**
    - **What is benchmarking?** Benchmarking compares your organisation's energy performance to industry standards or similar facilities. It provides information about areas where improvements can be made.

- **Why is it important?** Benchmarking identifies inefficiencies, establishes performance goals, and enables informed decision-making.
- **How to do it?** Collect data on energy consumption, compare it to relevant benchmarks, and identify gaps.

2. **Reporting requirements:**
    - **What are reporting requirements?** These are mandatory disclosures about energy usage and efficiency measures. They vary according to region and industry.
    - **Why is it important?** Reporting promotes transparency, accountability, and energy-efficient practices.
    - **How to comply?** Understand local regulations, monitor energy consumption, and provide accurate reports.
3. **Strategies for ensuring compliance:**
    - **Energy audits**: Regular energy audits help to identify areas for improvement. Hire professionals to evaluate your facility's energy usage.
    - **Invest in efficiency measures**: Use energy-efficient technologies like LED lighting, smart HVAC systems, and insulation.
    - **Employee training**: Educate employees about energy-saving practices.
    - **Policy implementation**: Create internal policies that comply with regulations.
    - **Monitoring and verification:** Continuously monitor energy usage and ensure compliance.

The role of competition in energy supply for effective energy management and conservation

Competition in energy supply is increasingly seen as critical to improving energy management and conservation. This section investigates how competitive dynamics in the energy market can lead to increased energy efficiency, drive innovation in sustainable technologies, and eventually contribute to more sustainable energy management practices.

Understanding energy market structures

Understanding various market structures is critical for understanding the impact of competition on energy management and conservation.

1. Monopolistic markets: Historically, energy markets were dominated by monopolies, with a single supplier controlling the entire value chain. This structure frequently resulted in inefficiencies and higher prices due to a lack of competitive pressure.
2. Deregulated markets: Many regions have adopted deregulation, which separates energy production and distribution. In these markets, multiple suppliers compete for customers, potentially leading to lower prices and better service offerings.

Benefits of competitive energy markets

The introduction of competition in energy supply brings several benefits, which can directly impact energy management and conservation.

1. Cost efficiency: Competitive markets drive down prices as suppliers strive to reduce operating costs and improve production efficiency in order to attract and retain customers.
2. Innovation and technological advancement: To stand out in a competitive market, businesses invest in cutting-edge energy technologies such as renewable energy and energy-efficient products and services.
3. Improved customer choice: In competitive markets, consumers can select suppliers based on price, service quality, and sustainability practices, encouraging suppliers to prioritise energy efficiency and conservation in their offerings.
4. Enhanced energy efficiency services: To differentiate themselves in a competitive market, suppliers may offer energy efficiency programmes, audits, and conservation tips to help consumers use energy more efficiently.

9.3 Case Studies: Impact of Competition on Energy Conservation

Several case studies illustrate how competition can drive energy conservation:

Case study: Deregulation and smart energy management in the European Union

Background

The European Union (EU) faced multiple challenges related to energy security, affordability, and decarbonisation. The 2022 energy crisis, triggered by Russia's invasion of Ukraine, intensified concerns about high energy prices and supply disruptions. As a response, the EU embarked on reforms to redesign its electricity market.

Key observations

1. **Deregulation and market transformation**:
   - **Energy crisis context**: The 2022 energy crisis prompted discussions on redesigning the EU's electricity market.
   - **Short-term measures**: The REPowerEU plan aimed to phase out Russian fossil fuel imports, diversify supplies, boost energy savings, and accelerate the clean energy transition.
   - **Long-term structural reforms**: The EU sought to make the market more resilient, reduce price volatility, and ensure secure energy supplies from clean sources.
2. **Smart energy management adoption**:
   - **Competitive packages**: Energy suppliers bundled smart energy solutions with their services, enticing consumers with competitive pricing.
   - **Awareness and education**: Consumers became more aware of energy efficiency, leading to increased adoption of smart metres, programmable thermostats, and connected appliances.
   - **Regulatory support**: Favourable policies encouraged investment in energy-efficient technologies.
3. **Residential and commercial sectors**:
   - **Residential buildings**: Smart metres allowed homeowners to monitor and control energy consumption remotely. Demand response strategies helped adjust usage during peak hours.
   - **Commercial establishments**: Smart HVAC systems, lighting controls, and predictive maintenance technologies optimised energy usage in commercial buildings.

Implications

- **Environmental impact**: Increased adoption of smart energy management systems contributed to reduced carbon emissions and a more sustainable energy landscape.
- **Economic benefits**: Consumers enjoyed cost savings, while suppliers gained a competitive edge.
- **Technological advancements**: Innovation in smart grid infrastructure, data analytics, and IoT devices accelerated.

Specific example: Smart metre penetration

A historic milestone was reached in the European energy sector as the penetration of *smart electricity metering* surpassed 50%. This adoption trend has been tracked since 2004, reflecting the EU's commitment to energy efficiency and technological advancement.

Conclusion

The EU's electricity market reforms aimed to create a better-connected, resilient, and consumer-oriented energy landscape. While this case study provides an illustrative example, actual outcomes may vary based on regional context and specific policies.

Case study: Renewable energy projects in the United States

Background

In the United States, several states have embraced competitive energy markets, allowing multiple suppliers to compete for consumers' business. This deregulation fosters innovation and encourages suppliers to differentiate themselves through various offerings, including renewable energy solutions.

Key observations

1. **State-level competition**:
    - **Market dynamics**: In states with competitive energy markets, suppliers vie for customers by providing diverse energy options.
    - **Environmental awareness**: Consumers increasingly prioritise sustainability and seek environmentally friendly energy sources.

2. **Renewable energy projects**:
    - **Solar farms**: Developers establish large-scale solar farms to harness sunlight and generate clean electricity. These projects contribute significantly to the state's renewable energy portfolio.
    - **Wind turbines**: Wind energy projects involve erecting wind turbines in suitable regions. These turbines convert wind power into electricity, reducing reliance on fossil fuels.
    - **Hydropower installations**: Utilising rivers and water bodies, hydropower projects generate electricity through turbines. Existing dams can be retrofitted for hydropower production.
    - **Biomass facilities**Facilities: Biomass energy projects utilise organic materials (such as wood, agricultural residues, or waste) to produce heat or electricity.
    - **Geothermal power plants**: Geothermal energy taps into the earth's internal heat. Wells drilled into hot rock formations provide a sustainable energy source.
3. **Supplier strategies**:
    - **Green tariffs**: Suppliers offer green energy plans, allowing consumers to choose renewable sources. These tariffs often include certificates verifying the origin of the energy.
    - **Community solar**: Consumers participate in shared solar projects, even if they don't have solar panels on their property. This model promotes community engagement and renewable adoption.
    - **Corporate renewable agreements**: Large companies commit to purchasing renewable energy directly from projects. This benefits both the environment and the company's sustainability goals.

Implications

- **Environmental impact**: The rise in renewable energy projects contributes to reduced greenhouse gas emissions and a cleaner energy mix.
- **Consumer choice**: Consumers benefit from diverse energy options, aligning with their values and preferences.
- **Economic growth**: Renewable projects create jobs, stimulate local economies, and attract investment.

Specific examples

1. **California**:
   - **Solar power**: California leads in solar installations, with vast solar farms across the state.
   - **Wind farms**: Wind turbines dot the landscape, especially in regions like the Altamont Pass and Tehachapi Mountains.
   - **Community solar programs**: Californians participate in community solar initiatives, promoting shared renewable energy.
2. **Texas**:
   - **Wind energy**: Texas boasts extensive wind farms, particularly in West Texas and the Panhandle.
   - **Corporate PPAs**: Major companies, including tech giants, sign power purchase agreements (PPAs) directly with wind and solar projects.

Conclusion

In competitive U.S. states, the synergy between supplier strategies and consumer demand drives the growth of renewable energy projects. As more consumers prioritise sustainability, the transition to cleaner energy sources accelerates, benefiting both the environment and the economy.

For further investment opportunities in the U.S. renewable energy sector, consider reaching out to *SelectUSA*, the U.S. Department of Commerce's investment promotion initiative.

Regulatory frameworks and policies

While competition can drive energy conservation, appropriate regulatory frameworks and policies are required to support these market structures.

1. Consumer protection laws: Ensure that consumers are protected from the negative effects of competition, such as misleading marketing or unfair contract terms.
2. Environmental regulations: Set standards requiring energy suppliers to integrate renewable energy and conservation practices into their operations.
3. Provide subsidies or tax breaks to suppliers who invest in renewable energy sources or energy-efficient technologies.

Challenges and considerations

Despite the benefits, competition in energy supply poses several challenges:

1. Market power and manipulation: Even in the absence of strict regulations, dominant suppliers may continue to exert undue influence on the market, undermining the benefits of competition.
2. Infrastructure and investment: Moving to a competitive market may necessitate significant investment in energy infrastructure, which can be a barrier in some areas.
3. Consumer awareness and education: In order to make informed decisions in a competitive market, consumers must have access to adequate information and resources.

References

[1] Wikler.G. et al., Best practices in energy efficiency and load management programs, 1016383, EPRI, Palo Alto, CA, 2008.

[2] IEA (2023), Energy Efficiency 2023, IEA, Paris https://www.iea.org/reports/energy-efficiency-2023, Licence: CC BY 4.0. (Accessed: 08 May 2024)

[3] IEA (2024), Global EV Outlook 2024, IEA, Paris https://www.iea.org/reports/global-ev-outlook-2024, Licence: CC BY 4.0. (Accessed: 08 May 2024)

[4] IEA (2023), World Energy Outlook 2023, IEA, Paris https://www.iea.org/reports/world-energy-outlook-2023, Licence: CC BY 4.0 (report); CC BY NC SA 4.0 (Annex A). (Accessed: 08 May 2024)

[5] Elegbede, O. and Tippett, A. (2022) *Understanding the U.S. Renewable Energy Market: A Guide for International Investors 2022*. Available at: https://www.trade.gov/sites/default/files/2022-04/2022SelectUSARenewableEnergyGuide.pdf. (Accessed: 08 May 2024)

[6] *Case Studies of Renewable Thermal Energy, Centre for Climate and Energy Solutions*. Available at: https://www.c2es.org/document/case-studies-of-renewable-thermal-energy/ (Accessed: 8 May 2024).

[7] *Case Studies of Market Transformation: Energy Efficiency and Renewable Energy*. Available at: https://www.un.org/esa/sustdev/sdissues/energy/publications&reports/market_tranformation.pdf (Accessed: 08 May 2024).

[8] Elegbede, O. and Tippett, A. (2022) *Understanding the U.S. Renewable Energy Market: A Guide for International Investors 2022*. Available at:

https://www.trade.gov/sites/default/files/2022-04/2022SelectUSARenewableEnergyGuide.pdf. (Accessed: 08 May 2024)

[9] How US states can advance a successful clean-energy transition | McKinsey, www.mckinsey.com. Available at: https://www.mckinsey.com/industries/public-sector/our-insights/how-us-states-can-advance-a-successful-clean-energy-transition. (Accessed: 08 May 2024)

[10] Vidyakar, V. Top 35 Projects Based on Renewable Energy, www.skyfilabs.com. Available at: https://www.skyfilabs.com/project-ideas/latest-projects-based-on-renewable-energy (Accessed: 10 May 2024).

[11] Renewable Energy Market Size, Share, Competitive Landscape and Trend Analysis Report by Type and End Use : Global Opportunity Analysis and Industry Forecast, Allied Market Research, Alliedmarketresearch.com. Available at: https://www.alliedmarketresearch.com/renewable-energy-market. (Accessed: 08 May 2024)

[12] Europe Smart Electricity & Gas Metering Market Report 2021: EU Energy Policies Driving the Adoption of Smart Metering and the Latest Market Developments - ResearchAndMarkets.com, https://www.businesswire.com/news/home/20211103005759/en/Europe-Smart-Electricity-Gas-Metering-Market-Report-2021-EU-Energy-Policies-Driving-the-Adoption-of-Smart-Metering-and-the-Latest-Market-Developments---ResearchAndMarkets.com (Accessed: 08 May 2024)

[13] Reforming the EU electricity market (2023), EPRS | European Parliamentary Research Service, European Union, https://www.europarl.europa.eu/RegData/etudes/BRIE/2023/739374/EPRS_BRI%282023%29739374_EN.pdf

[14] Rules for Europe's Electricity Market, European Commission, https://energy.ec.europa.eu/system/files/2019-04/electricity_market_factsheet_0.pdf (Accessed: 10 May 2024).

[15] Energy Deregulation Around the World: A Comprehensive Guide. www.electricchoice.com. Available at: https://www.electricchoice.com/blog/energy-deregulation-world/. (Accessed: 08 May 2024)

10

Energy Management Tools and Software

Chapter Preview

In this chapter, we briefly review energy management tools and software, both commercial and open source. We review the key features in such software, as well as data analysis and monitoring in energy management.

10.1 Energy Management Software Solutions

Energy management software has emerged as a critical tool for energy conservation. These solutions allow professionals to effectively track, analyse, and optimise their energy consumption. This section will look at a variety of commercial and open-source software tools that cater to different organisational needs and scales, highlighting how they help to improve energy management strategies.

Key features of energy management software

Energy management software typically includes several key features:

- Real-time monitoring: Enables continuous observation of energy consumption, allowing for the early detection of anomalies or excess usage.
- Data analysis and reporting: These tools examine consumption patterns, generate reports, and offer actionable insights.
- Forecasting: Software often includes the ability to forecast future energy needs based on historical data, which helps with planning and decision-making. These features are critical for identifying areas of inefficiency, facilitating targeted improvements, and forecasting future energy needs.

Energy management software is essential for optimising energy consumption, lowering costs, and promoting sustainability. Let us look at some important aspects of energy management software solutions:

1. **Monitoring and real-time insights**:
   - Energy management software allows for real-time monitoring of energy usage from multiple sources. It is useful for tracking consumption patterns, identifying peak hours, and detecting anomalies.
   - With accurate data, professionals can make informed decisions to improve energy distribution and reduce waste.
2. **Energy data analytics**:
   - These tools use historical data to identify trends, patterns, and inefficiencies.
   - Professionals can gain insight into which areas use the most energy, allowing for targeted improvements.
3. **HVAC systems controls**:
   - Energy management software works with HVAC systems to control temperature, ventilation, and air conditioning.
   - Smart controls ensure that operations run smoothly while keeping users comfortable.
4. **Carbon and sustainability reporting**:
   - Professionals can compile reports on carbon emissions, energy efficiency, and sustainability initiatives.
   - Accurate reporting facilitates compliance with environmental standards.
5. **Commercial energy management software**:
   - **Metasys Building Automation System**:
     - **Description**: Metasys is a world-class technology system designed for energy management efficiency. It connects commercial HVAC, lighting, and security systems.
     - **Features**:
       - Real-time energy usage monitoring
       - Integration with various devices and services
       - Control algorithms for efficient energy distribution
   - **IBM Envizi ESG Suite**:
     - **Description**: The IBM Envizi ESG Suite is a comprehensive data and analytics software. It collects, manages, and derives insights from environmental, social, and governance aspects.

- ○ **Features**:
  - ▪ Real-time monitoring
  - ▪ Data analytics for informed decision-making
  - ▪ Sustainability reporting

6. **Open-source energy management software**:
   - **OpenEMS (Open energy management system)**:
     - ○ **Description**: OpenEMS is a modular platform for energy management applications. It caters to monitoring, controlling, and integrating energy storage, renewable energy sources, and complementary devices.
     - ○ **Features**:
       - ▪ Real-time monitoring and control
       - ▪ Integration with renewable energy sources
       - ▪ Device-independent control algorithms
     - ○ **Source**: OpenEMS Introduction
     - ○ **GitHub Repository**: OpenEMS on GitHub
   - **Building Energy Management Open-Source Software (BEMOSS)**:
     - ○ **Description**: BEMOSS is engineered to improve the sensing and control of equipment in small- and medium-sized commercial buildings that lack building automation systems.
     - ○ **Features**:
       - ▪ Monitoring and optimisation of energy usage
       - ▪ Cost reduction and sustainability promotion
     - ○ **Source**: BEMOSS Overview

10.2 Data Analysis and Monitoring in Energy Management

Data analysis and monitoring are critical components in the quest for efficient energy management. They provide the fundamental insights required to make sound energy-related decisions. This section looks at the technologies and methodologies used to collect and analyse energy data, as well as how this information is used to develop energy-saving strategies that improve overall efficiency.

Technologies for energy data collection

A range of technologies has revolutionised the collection of energy data:

- Smart metres: These devices record energy usage in greater detail than traditional metres, providing useful data for analysis.
- Sensors and IoT devices: Sensors located throughout an energy system can collect granular data, which can then be integrated by IoT devices for comprehensive monitoring. This technology enables the collection of detailed and accurate energy usage data, which serves as the foundation for effective energy management strategies.

Real-time monitoring systems

Real-time monitoring systems are critical for timely insights into energy consumption patterns:

- Immediate feedback: These systems provide real-time data on energy usage, allowing for the rapid identification of inefficiencies or spikes in energy consumption.

Analysing energy data for insights

Analysing collected energy data is crucial for understanding consumption patterns and identifying opportunities to save.

- Data Analysis tools: Software tools and platforms are used to process and analyse energy data. This could include AI-powered models or predictive analytics that forecast future energy consumption and identify trends.
- Informing decisions: Data analysis insights inform strategic energy conservation decisions, such as optimising equipment usage or changing operational procedures.

Enhancing energy efficiency with data-driven strategies

Data-driven strategies are at the heart of enhancing energy efficiency:

- Setting goals and measuring progress: Data can help you set realistic energy-saving targets and track your progress towards them.
- Organisations can develop tailored energy-saving strategies based on their specific data insights, resulting in more effective and sustainable energy management.

Understanding consumption patterns, identifying inefficiencies, and making informed decisions are all critical components of effective energy management. Here's how data analytics and monitoring play an important role:

1. **Real-time monitoring**:
    - **Purpose**: Real-time monitoring systems continuously monitor energy consumption across multiple processes and equipment.
    - **Benefits**:
        - Immediate detection of anomalies or deviations.
        - Details about peak energy demand hours.
        - Early detection of potential problems.
2. **Energy data analytics**:
    - **Purpose**: Analysing historical data to uncover trends, patterns, and inefficiencies.
    - **Benefits**:
        - Identifying areas with high energy consumption.
        - Identifying opportunities for improvement.
        - Encourages evidence-based decision-making.
3. **Granular insights**:
    - **Purpose**: Detailed data granularity enables for targeted actions.
    - **Examples**:
        - Determine which equipment consumes the most energy.
        - Identifying specific processes with high energy demands.
4. **Predictive analytics**:
    - **Purpose**: Forecasting future energy needs based on historical data.
    - **Benefits**:
        - Optimising energy distribution.
        - Planning for peak demand periods.
        - Reducing costs through efficient resource allocation.
5. **Integration with IoT**:
    - **Purpose**: Leveraging Internet of things (IoT) sensors for real-time data collection.
    - **Benefits**:
        - Enhanced accuracy and timeliness.

- ○ Improved visibility into energy usage patterns.

6. **Energy efficiency strategies**:
   - **Purpose**: Data-driven insights inform strategies for reducing consumption.
   - **Examples**:
     - ○ Adjusting the HVAC settings during off-peak hours.
     - ○ Implementing load shedding during peak demand periods.
     - ○ Identifying potential for renewable energy integration.

10.3 Implementing Energy Management Systems

Implementing an energy management system (EMS) can be transformative for businesses looking to improve their energy efficiency and cut costs. This journey, from initial planning to continuous improvement, necessitates thoughtful consideration and strategic execution. This section walks you through the different stages of implementing an EMS, highlighting key steps and addressing potential challenges.

Initial planning and assessment

The foundation of a successful EMS implementation lies in thorough initial planning:

- Defining objectives: Clearly state what the organisation hopes to achieve with the EMS, whether it is to reduce energy consumption, cut costs, or improve sustainability.
- Energy audit: Conduct a thorough energy audit to better understand your current energy usage and identify areas for improvement. This audit will help shape the scope and specifications of the EMS.

Selecting the right energy management system

Choosing the right EMS is critical:

- Compatibility and scalability: Determine whether the system is compatible with existing infrastructure and scalable for future needs.
- Features and functionality: Evaluate systems based on their features, such as real-time monitoring, data analytics, and reporting.
- Industry-specific needs: Consider any industry-specific requirements that may influence system selection.

Deployment and integration

Deploying the EMS involves several key steps:

- Installation: Install the system's software and hardware components, making sure they are properly configured and integrated.
- Staff training: Provide comprehensive training to staff and stakeholders to familiarise them with the system's features and benefits.
- Process integration: Easily integrate the EMS into existing organisational processes and workflows.

Continuous monitoring and improvement

An EMS is not a set-and-forget solution:

- Ongoing monitoring: Continuously monitor the system's performance to ensure it meets the energy targets.
- Regular system reviews and updates: Conduct regular system reviews and implement updates or adjustments in response to performance data and changing organisational needs.

Addressing challenges in implementation

Implementation can come with challenges:

- To overcome resistance to change, highlight the benefits of the EMS and involve employees in the implementation process.
- Technical issues: Address technical challenges by collaborating with the system provider and seeking expert advice.
- Budget constraints: If money is a concern, consider a phased implementation strategy that prioritises key areas first.

References

[1] *Best Energy Management Software in 2019 | G2*. Available at: https://www.g2.com/categories/energy-management (Accessed: 08 May 2024).

[2] *Best Energy Management Software 2024 Capterra*. Available at: https://www.capterra.com/energy-management-software/ (Accessed: 08 May 2024).

[3] *Top 23 Energy Management Software Solutions for Sustainable Operations* (2024) *www.process.st*. Available at: https://www.process.st/energy-management-software/ (Accessed: 8 May 2024).

[4] *Networking and knowledge: Why conferences are vital for renewable energy leaders - Tamarindo* (2023). Available at: https://tamarindo.global/news/networking-and-knowledge-why-conferences-are-vital-for-renewable-energy-leaders/ (Accessed: 8 May 2024).

[5] *Building Energy Management Open-Source Software (BEMOSS) Energy.gov*. Available at: https://www.energy.gov/eere/buildings/articles/building-energy-management-open-source-software-bemoss. (Accessed: 08 May 2024)

[6] *Climate resilience – Power Systems in Transition – Analysis IEA*. Available at: https://www.iea.org/reports/power-systems-in-transition/climate-resilience. (Accessed: 08 May 2024)

[7] *Making the energy sector more resilient to climate change – Analysis , IEA*. Available at: https://www.iea.org/reports/making-the-energy-sector-more-resilient-to-climate-change. (Accessed: 08 May 2024)

[8] *Energy end-use data collection methodologies and the emerging role of digital technologies – Analysis IEA*. Available at: https://www.iea.org/reports/energy-end-use-data-collection-methodologies-and-the-emerging-role-of-digital-technologies. (Accessed: 08 May 2024)

[9] *Practical ways to apply data analytics and AI for energy management and emission control in the steel and cement industries* (2024) *Industrial Software*. Available at: https://new.abb.com/industrial-software/digital/energy-managment/practical-ways-to-apply-data-analytics-and-ai-for-energy-management-and-emission-control-in-the-steel-and-cement-industries. (Accessed: 08 May 2024)

[10] *Inclusive and Sustainable Industrial Development Practical Guide for Implementing an Energy Management System*. Available at: https://www.unido.org/sites/default/files/2017-11/IEE_EnMS_Practical_Guide.pdf. (Accessed: 08 May 2024)

[11] *Implementing Energy Management Systems - Better Buildings Initiative*. Available at: https://betterbuildingssolutioncenter.energy.gov/sites/default/files/Implementing_Energy_Management_Systems.pdf (Accessed: 08 May 2024).

[12] *Best Energy Management Software in 2023 [UPDATED]* (2022) *Facilio Blog*. Available at: https://facilio.com/blog/energy-management-software/.

[13] Rao, M. (2024) *Best Energy Management Software in 2024 [updated]*, *Facilio Blog*. Available at: https://facilio.com/blog/energy-management-software/ (Accessed: 08 May 2024).

11

Financing Energy Efficiency Projects

Chapter Preview

In this chapter, we review financing energy efficiency projects. We review associated concepts such as payback period, net present value. We also delve into funding options for energy efficiency projects.

11.1 Cost Analysis

11.1.1 Simple payback period

The simple payback period (SPB) is an important financial metric for determining the efficiency of an investment, particularly in the context of energy projects and other capital-intensive initiatives. It provides a simple way to compare different projects by calculating the time required to recoup the initial investment costs through future savings or earnings. Despite its simplicity and widespread use, SPB has limitations and areas where it excels or falls short.

Definition and calculation of SPB

The simple payback period is the number of years required to recover the project cost from the net benefits (such as savings or revenue) generated by the investment. The formula for calculating SPB is fairly straightforward (eqn (11.1)):

$$\text{SPB(Years)} = \frac{\text{Capital investment}}{\text{Annual savings}}, \tag{11.1}$$

Here:

- Capital investment refers to the total upfront cost required to fund a project.
- Annual savings refers to the annual financial benefits generated by the investment, which do not take into account any value changes over time.

Strengths of using SPB

1. Ease of use: SPB's main advantage is its simplicity. It gives a quick overview of the investment recovery timeline that is simple to calculate and understand.
2. Risk assessment: SPB is especially useful for evaluating projects that involve significant uncertainties or risks. It enables investors to estimate the time period during which their capital will be 'at risk' before recovering through project returns. This can be critical for decision-making in environments with high volatility or uncertain returns.

Limitations of SPB

While SPB is a valuable tool, it has several limitations that may reduce its utility in more complex financial analyses:

1. One of the most significant disadvantages of the SPB method is that it ignores the time value of money (TVM). TVM is the idea that money available now is more valuable than the same amount in the future because of its potential earning capacity. Due to this limitation, SPB may not always accurately reflect an investment's true profitability or risk.
2. SPB is not appropriate for projects with complex financial structures because financing and tax implications are critical. SPB's simplicity can be a disadvantage because it fails to capture the nuances introduced by financing arrangements or tax incentives, which can significantly alter a project's financial landscape.
3. SPB is not suitable for mutually exclusive projects because it can be misleading when choosing one option over the others. This method fails to account for differences in investment scale, which can skew the comparative analysis. A project with a shorter payback period may require less investment and potentially provide less overall benefit than a larger project with a longer payback period.

The simple payback period is a useful, if basic, tool for financial analysis. It works well in situations requiring quick, straightforward assessments, such as determining an investment's risk exposure period. However, given its limitations, it should be used with caution, especially for complex investments or when precise, comprehensive financial analysis is required. Investors should supplement SPB with other financial metrics that account for the time value of money, such as net present value (NPV) or internal rate of return (IRR), to get a more complete picture of an investment's prospects.

11.1.2 Discounted payback period: An in-depth analysis

The discounted payback period (DPP) is a more advanced financial metric that addresses one of the primary limitations of the simple payback period (SPB) by incorporating the time value of money (TVM) into its calculations. This change makes DPP a more precise tool for assessing the profitability and risk of an investment, especially in scenarios where cash flows span longer time periods.

Understanding discounted payback period

The discounted payback period is the time required to recover the cost of an investment while accounting for the declining value of future cash flows. Unlike SPB, which assumes that a dollar today will have the same value in the future, DPP adjusts each future cash flow to its current value before adding them together to calculate the payback period. This method more accurately represents the investment's financial viability and risk.

Calculation of DPP

The DPP is calculated by:

1. Discounting the cash flows: Each cash flow generated by the investment over time is discounted to its present value using a specific discount rate. This rate could represent the cost of capital, the desired rate of return, or any other rate that reflects the investment's opportunity cost.
2. Summing the discounted cash flows: The discounted cash flows are added together until they equal or exceed the original capital investment. This occurs during the discounted payback period.

The formula to find the present value (PV) of each cash flow is given by eqn (11.2):

$$PV = \frac{CF}{(1+r)^t}, \tag{11.2}$$

where

- *CF* is the cash flow in year *t*;
- *r* is the discount rate;
- *t* is the time in years from initial investment.

Benefits of using DPP

1. Incorporating time value of money: DPP provides a more realistic assessment of an investment's returns by acknowledging that money earned in the future is less valuable than money earned today.
2. Risk analysis: By calculating the time required to repay the investment using discounted cash flows, DPP provides a more complete picture of the investment's risk profile, which is especially useful for long-term projects.
3. Decision-making precision: DPP is especially useful in capital budgeting decisions, allowing managers to prioritise projects that not only repay their initial investment quickly but also generate returns that are more consistent with the firm's financial goals.

Limitations of DPP

While DPP offers a more nuanced approach than SPB, it is not without its drawbacks:

1. DPP is more complex and time-consuming to compute than SPB because it requires calculating the present value of each future cash flow.
2. Subjectivity: The discount rate chosen can have a significant impact on the DPP, and this rate can sometimes be subjective, depending on assumptions about the project's cost of capital or risk profile.
3. Ignores cash flows. Post-payback: Similar to SPB, DPP does not account for cash flows that occur after the payback period, potentially overlooking the total value generated by the investment.

Practical applications

DPP is especially useful in industries that require large investments with long-term financial returns, such as real estate development, large-scale manufacturing, and renewable energy projects. It enables financial analysts and decision-makers to determine which projects are most aligned with their organisations' financial stability and long-term objectives.

The discounted payback period builds on the simple payback period by incorporating the critical financial principle of time value of money, allowing for a more rigorous evaluation of investment projects. While it adds complexity to the analysis, the benefits of using DPP to assess the true financial impact of investment decisions make it an important tool in the financial analyst's toolbox, particularly for long-term and capital-intensive projects.

11.1.3 Net present value (NPV): A comprehensive overview

Net present value (NPV) is an important financial metric for determining the profitability and viability of an investment project. By calculating the sum of discounted cash flows over the life of the project, NPV provides a comprehensive view of the project's potential to generate value, taking into account both costs (cash outflows) and revenues (cash inflows). NPV is highly valued because it can be applied to a wide range of investment scenarios, including mutually exclusive projects and social cost assessments.

Calculating NPV

NPV is calculated by discounting all future cash flows to their present value and then summing the results. The formula for net present value is given by eqn (11.3):

$$\text{NPV} = \sum_{t=0}^{n} \frac{C_t}{(1+r)^t}, \tag{11.3}$$

where

- C_t is the net cash inflow during the period t;
- r is the discount rate; and
- n is the total number of periods.

The discount factor used in the NPV calculation is expressed as $\frac{1}{(1+r)^n}$, where r represents the interest rate and n denotes the number of years used to calculate the discount factor. This factor adjusts future cash flows to reflect the time value of money, recognising that money received in the future is less valuable than money received now.

Advantages of using NPV

1. Comprehensive evaluation: NPV considers all cash flows of the project, providing a comprehensive picture of its financial implications.
2. Time value of money: By incorporating the discount rate into the calculation, NPV can account for the declining value of future cash flows, making it a more accurate measure of a project's profitability.
3. Flexibility: NPV can be applied in a variety of scenarios, ranging from simple project evaluations to complex capital budgeting for mutually exclusive projects. It can handle varying cash flows over different time periods, which is critical in real-world scenarios.

Application in investment decisions

NPV is highly recommended for evaluating investment options, especially in cases such as:

- Mutually exclusive projects: When deciding between options where only one project can be chosen, NPV provides a solid foundation for comparison by reflecting the size of cash flows and the scale of investment. This is critical because it allows investors to evaluate the absolute amount of value created by each option, rather than just percentage returns.
- Social costs: NPV is best suited for projects with significant social implications, where the benefits or costs go beyond simple cash transactions. This includes environmental projects, public infrastructure, and social programmes, all of which can provide widespread benefits over time.

NPV and societal implications

When evaluating projects from a societal perspective, the net present value (NPV) is critical because it quantifies the economic value of social benefits and costs. For example, in environmental projects, NPV can be used to calculate the long-term savings from reduced pollution versus the initial costs of implementing green technology.

Challenges and considerations

Despite its benefits, NPV calculation is heavily reliant on the accuracy of inputs such as the discount rate and future cash flow projections. These elements can be highly speculative and uncertain, especially in long-term projects where market conditions and economic environments can shift.

Net present value is a powerful tool for financial analysis, providing a thorough and nuanced understanding of investment opportunities. It is especially useful in situations where the size of the investment and long-term effects are important considerations. NPV allows businesses and policymakers to make informed decisions that not only improve economic efficiency but also benefit society. Understanding and applying NPV effectively can lead to better investment decisions and more efficient resource allocation in both the private and public sectors.

11.2 Financing Energy Efficiency Projects

11.2.1 Funding options for energy conservation

This section discusses the various funding options for energy conservation projects, such as grants, loans, and incentives. We will go over the eligibility requirements and procedures for obtaining these funds, as well as the benefits and drawbacks of each funding source. This information is intended to help professionals identify and obtain financial support for their energy efficiency initiatives.

Financing energy efficiency projects

Obtaining adequate financing is a critical step in the implementation of energy-efficiency projects. The right funding can transform ambitious plans into tangible results, whether for infrastructure upgrades, the implementation of new technologies, or research. This section delves into a variety of funding sources, from grants and loans to incentives and rebates, each with its own criteria, processes, and set of challenges and opportunities. Understanding these options is critical for professionals who want to navigate the financial landscape of energy conservation successfully.

Grants: Free funding with specific requirements

Grants are an attractive funding option because they do not require repayment. Government agencies, non-governmental organisations, and private foundations frequently provide funding for energy conservation initiatives. However, getting a grant is extremely competitive. Applicants must meet specific eligibility criteria, which are typically aligned with the grantor's goals. The application process usually includes a detailed proposal outlining the project's scope, expected outcomes, and environmental impact. While grants are a great source of funding, they frequently come with strict reporting requirements and may only cover a portion of the total project costs.

Loans: Borrowed capital with terms and conditions

Loans are a common source of funding that must be repaid with interest over a set period. They can be obtained from banks, specialised green financing institutions, or government-sponsored programmes. Securing a loan requires demonstrating the project's financial viability. This entails presenting a strong business case with detailed financial projections. Interest rates and terms can vary greatly, so applicants must carefully consider the long-term financial implications, such as how loan repayments will be managed within the project's budget.

Incentives and rebates: Encouraging energy efficiency

Incentives and rebates are financial rewards provided by governments and private entities to encourage the use of energy-efficient practices. These could include tax credits for implementing renewable energy solutions, rebates for purchasing energy-efficient equipment, or lower tariffs for using renewable energy. To qualify, projects must often meet specific criteria, such as demonstrating energy savings or utilising specific technologies. While these incentives can significantly reduce a project's overall cost, understanding the eligibility requirements and application process can be difficult.

Navigating the funding landscape: Strategies and tips

Finding and securing the appropriate funding necessitates a strategic approach. Professionals should begin by clearly understanding their project's requirements and then match them with the most appropriate funding source. Preparing a compelling application is critical; it should include a solid project plan, realistic goals, and a clear demonstration of the expected ROI. Networking with industry peers and attending relevant workshops can provide useful information about the funding landscape. Maintaining financial transparency and leveraging partnerships can also help to increase the proposal's credibility and appeal.

Financing as a stepping stone to energy conservation

Financing is critical to the successful execution of energy efficiency projects. A thorough understanding of the various funding options, combined with a strategic approach to obtaining this funding, can significantly increase the likelihood of a project's success. Professionals in this field are encouraged to explore all available financing options for their initiatives, using them as a stepping stone towards their energy conservation goals.

Financing energy efficiency projects. These mechanisms play a crucial role in supporting initiatives to reduce energy consumption, lower fossil fuel emissions, and enhance overall energy efficiency. Here are some key avenues:

1. **Energy Efficiency and Conservation Block Grant (EECBG) Programme**:
    - The *EECBG Programme*, administered by the *Department of Energy (DOE)*, targets states, local governments, and Tribes. Its

primary goal is to implement strategies that enhance energy efficiency.
- **Funding mechanism**: The program offers both *formula grants* (distributed based on specific criteria) and *competitive grants*.
- **Eligibility**: State, local government, and tribal entities.
- **Available funds**: The program has a total funding amount of *$550 million*.
- **Use cases**: EECBG funds can be utilised for a wide range of energy conservation projects.
- **Technical assistance**: The program provides technical assistance resources to support project implementation.
- **Recent developments**: In 2023, DOE announced twelve selectees for competitive EECBG grants1.

2. **On-bill financing programs**:
    - These programs allow for low-cost financing of energy efficiency and electrification measures. The costs are spread out as monthly utility bill line-items, with no upfront expenses.
    - **Eligibility**: Available for various entities, including non-profits.
    - **Benefits**: No need for significant initial capital; repayment occurs over time through utility bills.
3. **Green banks**:
    - Green banks specialise in providing low-cost financing for clean energy investments.
    - **Targeted investments**: They focus on supporting non-profits, rural areas, and low- to medium-income communities.
    - **Role**: Green banks bridge the gap between available financing and the requirements for sustainable energy projects.
4. **Voucher application option**:
    - Instead of traditional grants, local governments and Tribes eligible for EECBG formula funding can apply for technical assistance and/or equipment rebate vouchers.
    - **Benefits**: Vouchers offer flexibility and can be used for specific needs, such as technical support or equipment upgrades.

Cost-benefit analysis Analysis

A thorough cost-benefit analysis is critical for determining the feasibility and potential return on investment of energy efficiency projects. This section establishes a framework for such an analysis, covering topics such as initial costs, long-term savings, and environmental impact.

Securing grants and loans

Securing grants and loans is a critical step in the process of implementing any energy efficiency project. Despite its complexity, this process can be successfully navigated with careful planning and a strategic approach. Understanding the nuances of proposal writing, eligibility requirements, and the application process is critical. This section offers practical advice to help professionals navigate the complexities of applying for grants and loans, increasing their chances of obtaining the funding they require for their energy projects.

Preparing a compelling proposal

A compelling proposal is critical to obtaining funding. It should clearly outline the project objectives, provide a detailed description of the project's expected outcomes, and include a well-structured budget. Here are some tips for making an impactful proposal:

- Clarity and precision: Clearly articulate the project goals and how they relate to the funder's objectives.
- Highlight the project's environmental impact and sustainability benefits.
- Including data and case studies: Strengthen the proposal by including relevant data, pilot study results, or case studies to demonstrate feasibility and effectiveness.

Understanding and meeting eligibility criteria

Understanding the eligibility criteria established by funding organisations is critical. These criteria determine not only who can apply, but also the types of projects that are funded. To meet these criteria:

- Conduct thorough research to understand the specific requirements of each funding body.
- Alignment with goals: Ensure that your project's objectives are consistent with the funder's mission and goals.
- Seeking clarification: Do not hesitate to contact the funding body for advice or to attend their informational workshops for a better understanding.

Navigating the application process

The application process can differ significantly between funding bodies. However, there are some common elements and pitfalls to be aware of.

- Adherence to guidelines: Carefully follow the application guidelines provided by the funding body.
- Effective communication: Communicate clearly and professionally throughout the process.
- Understanding the review process: Learn about how applications are reviewed and what factors influence decision-making.

Post-application: Follow-up and next steps

After submitting an application, the process doesn't end there. Follow-up is important:

- Staying engaged: Maintain contact with the funding body to demonstrate your ongoing interest and commitment to the project.
- Providing additional information: Be prepared to provide any additional details or clarifications requested by the funders.
- Plan B: Always have a backup plan or alternative funding sources in case the application fails.

Securing funding as a key step to energy efficiency

Securing grants and loans is a critical but difficult aspect of implementing energy efficiency projects. A well-written application that is aligned with the funder's criteria and effectively communicated can significantly increase the chances of success. Persistence and adaptability, combined with a thorough understanding of the funding landscape, are essential for navigating this journey successfully.

11.2.2 Cost-benefit analysis for energy efficiency projects

1. **The importance of cost-benefit analysis:** A thorough cost-benefit analysis is required when determining the feasibility and potential return on investment of energy efficiency projects. Decision-makers can make more informed project implementation decisions by systematically assessing costs and benefits.
2. **Framework for cost-benefit analysis:** Here are the key components of a comprehensive cost-benefit analysis for energy efficiency initiatives:
    - **Upfront costs:**
        - Identify and quantify the initial project implementation costs. These could include equipment purchases, installation, labour, and administrative expenses.

  - If you need external funding, take into account any financing or loan fees.

- **Long-term savings:**
  - Estimate the long-term financial benefits of energy savings. These may include lower utility bills, maintenance costs, and operational expenses.
  - Consider the expected lifespan of energy-efficient measures, such as upgraded HVAC systems, lighting, and insulation.
- **Environmental impact:**
  - Assess the project's positive environmental effects. This could include cutting greenhouse gas emissions, reducing reliance on fossil fuels, and conserving natural resources.
  - Quantify the effects in terms of avoided emissions or other relevant metrics.
- **Payback period:**
  - Calculate the payback period, which is the time it takes for the cumulative savings to cover the initial investment.
  - A shorter payback period means a quicker return on investment.
- **Return on investment (ROI):**
  - Calculate ROI by comparing net benefits (savings minus costs) to the initial investment.
  - Calculate ROI as a percentage to determine the project's profitability.

3. **An hypothetical example:** Let's consider a hypothetical case study: The **London Residential Energy Efficiency Programme**. By analysing its costs and benefits, we can illustrate the process:
   - **Project description:**
     - The programme aims to retrofit residential buildings with energy-efficient appliances, insulation, and lighting.
     - Upfront costs include equipment purchase, installation, and administrative expenses.
   - **Benefits:**
     - Reduced electricity consumption leads to lower utility bills for homeowners.

- Improved insulation decreases the need for heating and cooling, resulting in long-term savings.
- Environmental benefits include reduced carbon emissions.

- **Cost-benefit outcome:**
  - The payback period is estimated at three years, indicating a relatively quick return on investment.
  - The ROI is calculated as 20%, demonstrating the project's financial viability.

4. **Additional resources:** For further insights into cost-effectiveness and quantifying benefits, refer to resources such as:
   - Understanding cost-effectiveness of energy efficiency programmes by the **Leadership Group of the National Action Plan for Energy Efficiency**.
   - Quantifying the multiple benefits of energy efficiency and renewable energy by the **U.S. Environmental Protection Agency (EPA)**.

Securing grants and loans

Securing grants and loans is a critical step in the process of implementing any energy efficiency project. This process, while often complex, can be successfully navigated with careful planning and strategic thinking. Understanding the nuances of proposal writing, eligibility requirements, and the application process is critical. This section offers practical advice to help professionals navigate the complexities of applying for grants and loans, increasing their chances of obtaining the funding they require for their energy projects.

Preparing a compelling proposal

A compelling proposal is critical to obtaining funding. It should clearly state the project's objectives, provide a detailed description, project expected outcomes, and present a well-structured budget. Here are some tips for making an impactful proposal:

- Clarity and precision: Clearly articulate the project goals and how they relate to the funder's objectives.
- Highlight the project's environmental impact and sustainability benefits.
- Including data and case studies: Strengthen the proposal by including relevant data, pilot study results, or case studies to demonstrate feasibility and effectiveness.

The process of financing energy efficiency projects, with a focus on securing grants and loans. Here's a comprehensive guide to assist you with this critical aspect:

1. **Understanding eligibility criteria**:
   - Before embarking on the grant and loan application process, it is critical to understand the eligibility requirements established by funding organisations. Each programme or opportunity may have unique requirements, so carefully review them.
   - For instance, the *Energy Efficiency and Conservation Block Grant (EECBG) Program* by the U.S. Department of Energy (DOE) offers funding for various energy-related activities. Eligible entities include states, territories, local governments, and tribes1. Verify your eligibility before proceeding.
2. **Research available funding opportunities**:
   - Familiarise yourself with the funding landscape. The *EERE Funding Opportunity eXCHANGE* serves as a hub for competitive solicitations (funding opportunity announcements or FOAs) throughout the year.
   - Keep an eye out for *requests for information (RFIs)* and *notices of intent (NOIs)* preceding FOAs. RFIs help identify needs, while NOIs provide insights into EERE's areas of interest.
3. **Prepare for application**:
   - Complete the necessary registrations:
     - Obtain a *Unique Entity ID (UEI number)* from SAM.gov.
     - Create a registered account with *Login.gov*.
     - Register in the *EERE Funding eXCHANGE* (ensure your email matches between Login.gov and eXCHANGE).
     - Consider creating a *Grants.gov account* for automatic FOA updates.
   - Understand the specific requirements for each FOA. Verify your eligibility and review the user manuals on eXCHANGE.
4. **Crafting a compelling grant proposal**:
   - **Introduction**:
     - Give an overview of your organisation's mission and past experience with renewable energy projects.

- **Problem statement**:
    - Clearly articulate the specific problem or challenge your project aims to solve within the renewable energy sector.
- **Project goals and objectives**:
    - Define your project's goals and measurable objectives.
- **Budget**:
    - Create a comprehensive budget, accounting for all project expenses.
- **Evaluation and measurement framework**:
    - Outline how you'll measure the success and impact of your project.
- **Expertise and sustainability**:
    - Highlight your team's expertise and emphasise the project's long-term sustainability.
- **Structure and key components**:
    - Ensure your proposal includes all necessary sections and follows the FOA guidelines.

5. **Application submission**:
    - Follow the specific process outlined in each FOA.
    - Verify your eligibility in Section IIIA of the FOA.
    - If selected, register with *FedConnect* to accept the award and receive award documents.

Securing funding requires persistence, thoroughness, and effective communication. Create a compelling proposal that not only piques funders' interest, but also communicates the urgency and significance of your renewable energy project. Best wishes on your journey to positive change!

References

[1] *EECBG PROGRAM Notice 23-01 Effective Date: April 25, 2023 Subject: Guidance for Eligibility of Activities Under the Energy Efficiency and Conservation Block Grant Program*. Available at: Phttps://www.energy.gov/scep/articles/energy-efficiency-and-conservation-block-grant-eligible-activities-and-program. (Accessed: 08 May 2024)

[2] *Funding application process | department of energy*. Available at: https://www.energy.gov/eere/funding/funding-application-process (Accessed: 08 May 2024).

[3] *A simple step-by-step guide for ngos on 'how to write proposals' - fundsforngos*. Available at: https://www2.fundsforngos.org/featured/a-simple-step-by-step-guide-for-ngos-on-how-to-write-proposals/ (Accessed: 08 May 2024).

[4] *Four tools for increasing sustainable energy finance, World Economic Forum*. Available at: https://www.weforum.org/agenda/2020/09/four-tools-for-increasing-sustainable-energy-financing/. (Accessed: 08 May 2024)

[5] EPA,OAR (2017) *Quantifying the Multiple Benefits of Energy Efficiency and Renewable Energy: A Guide for State and Local Governments | US EPA, US EPA*. Available at: https://www.epa.gov/statelocalenergy/quantifying-multiple-benefits-energy-efficiency-and-renewable-energy-guide-state. (Accessed: 08 May 2024)

[6] *Guidebook for Cost/Benefit Analysis of Smart Grid Demonstration Projects* . Available at: https://www.energy.gov/sites/prod/files/2017/01/f34/Guidebook-Cost-Benefit-Analysis-Smart-Grid-Demonstration-Projects.pdf. (Accessed: 08 May 2024)

[7] *Blueprint 5: Unlocking sustainable financing solutions for energy projects and programs with Revolving Loan Funds | Department of Energy*. Available at: https://www.energy.gov/scep/blueprint-5-unlocking-sustainable-financing-solutions-energy-projects-and-programs-revolving (Accessed: 08 May 2024).

[8] *Energy end-use data collection methodologies and the emerging role of digital technologies – Analysis IEA*. Available at: https://www.iea.org/reports/energy-end-use-data-collection-methodologies-and-the-emerging-role-of-digital-technologies. (Accessed: 08 May 2024)

[9] *Four tools for increasing sustainable energy finance World Economic Forum*. Available at: https://www.weforum.org/agenda/2020/09/four-tools-for-increasing-sustainable-energy-financing/. (Accessed: 08 May 2024)

[10] EPA,OAR (2017) *Quantifying the Multiple Benefits of Energy Efficiency and Renewable Energy: A Guide for State and Local Governments | US EPA, US EPA*. Available at: https://www.epa.gov/statelocalenergy/quantifying-multiple-benefits-energy-efficiency-and-renewable-energy-guide-state. (Accessed: 08 May 2024)

[11] *EERE Funding Opportunities, Energy.gov*. Available at: https://www.energy.gov/eere/funding/eere-funding-opportunities. (Accessed: 08 May 2024)

[12] *Financing Energy Efficiency Projects for Nonprofits | Article | EESI www.eesi.org*. Available at: https://www.eesi.org/articles/view/financing-energy-efficiency-projects-for-nonprofits (Accessed: 8 May 2024).

[13] Understanding Cost-Effectiveness of Energy Efficiency Programs: Best Practices, Technical Methods, and Emerging Issues for Policy-Makers A Resource of the National Action Plan for Energy Efficiency (2008). Available at: https://19january2017snapshot.epa.gov/sites/production/files/2015-08/documents/understanding_cost-effectiveness_of_energy_efficiency_programs_best_practices_technical_methods_and_emerging_issues_for_policy-makers.pdf. (Accessed: 08 May 2024)

[14] US EPA,OAR (2017) Quantifying the Multiple Benefits of Energy Efficiency and Renewable Energy: A Guide for State and Local Governments | US EPA, US EPA. Available at: https://www.epa.gov/statelocalenergy/quantifying-multiple-benefits-energy-efficiency-and-renewable-energy-guide-state. (Accessed: 08 May 2024)

12

Case Studies

Chapter Preview

This chapter presents a series of case studies from different industries and sectors where energy conservation measures have been successfully implemented. Each case study details the context, the energy-saving measures adopted, and the outcomes achieved.

12.1 Real-world Examples

Real-world examples of successful energy conservation and management across various sectors:

1. **Street smart lighting initiative (Indonesia)**:
   - **Context**: Indonesia's energy production primarily relies on coal, which is rising alongside economic growth.
   - **Energy-saving measures**:
     - **Street lighting upgrade**: Substituting conventional street lighting with more efficient technologies in cities and urban areas.
   - **Outcomes achieved**:
     - Reduced energy consumption for street lighting.
     - Improved urban lighting quality and safety.
     - Contributed to Indonesia's commitment to energy efficiency.
2. **Energy-efficient school buildings (United States)**:
   - **Context**: Many state schools in the US suffer from dated infrastructure and inefficient heating and cooling systems.
   - **Energy-saving measures**:

        - Retrofitting ageing school buildings with energy-efficient technologies.
        - Upgrading HVAC systems, insulation, and lighting.
    - **Outcomes achieved**:
        - Lower energy bills for schools.
        - Enhanced learning environments for students.
        - Reduced strain on the power grid.
3. **Low-carbon transport systems (Mexico)**:
    - **Context**: Transportation accounts for approximately half of energy consumption and 31% of global CO_2 emissions in Mexico.
    - **Energy-saving measures**:
        - Implementing rapid bus systems and promoting eco-driving practices.
    - **Outcomes achieved**:
        - Reduced air pollution in urban areas.
        - Increased efficiency in public transportation.
        - Contributed to Mexico's climate goals.
4. **Integrated energy conservation plans (various sectors)**:
    - **Context**: Integrated plans accommodate situational constraints and address multiple aspects of behaviour.
    - **Energy-saving measures**:
        - Evidence-based strategies tailored to specific contexts.
        - Adaptable approaches that consider technology advancements.
    - **Outcomes achieved**:
        - Holistic energy conservation across industries.
        - Cost savings, reduced emissions, and improved sustainability.

References

[1] Inspiring Examples of Energy Efficiency in Urban Environments, Unfccc.int. Available at: https://unfccc.int/news/inspiring-examples-of-energy-efficiency-in-urban-environments (Accessed: 10 May 2024). (Accessed: 08 May 2024)

[2] Here are 4 energy efficiency projects from around the world, World Economic Forum. Available at: https://www.weforum.org/agenda/2023/01/energy-efficiency-projects-innovations/. (Accessed: 08 May 2024)

[3] Attari, S.Z., Krantz, D.H. and Weber, E.U. (2016) 'Energy conservation goals: What people adopt, what they recommend, and why', Judgment and Decision Making, 11(4), pp. 342–351. Available at: https://doi.org/10.1017/s1930297500003776. (Accessed: 08 May 2024)

[4] Sustainable Energy Practices - Climate Action Planning (2019) Engineering LibreTexts. Available at: https://eng.libretexts.org/Bookshelves/Environmental_Engineering_%28Sustainability_and_Conservation%29/Book%3A_Sustainability_-_A_Comprehensive_Foundation_%28Cabezas%29/11%3A_Sustainable_Infrastructure/11.04%3A_Sustainable_Energy_Practices_-_Climate_Action_Planning. (Accessed: 08 May 2024)

13

Tips for Becoming an Energy-efficient Professional

Chapter Preview

In this chapter we discuss how individuals can advance personally and professionally in the field of energy conservation. It addresses topics such as continuous learning, certification programmes, and skill development. We will also talk about the importance of setting career goals that are consistent with sustainable and energy-efficient practices.

13.1 Continual Learning: Lifelong Commitment to Energy Conservation

In the ever-changing field of energy conservation, continuous learning is essential for professional success. Keeping up with the latest developments is critical. This requires regular engagement with new research, technological advancements, and policy changes. To stay current, professionals should subscribe to leading industry journals, follow relevant news outlets, and participate in online forums. Attending webinars, seminars, and workshops provides direct access to the latest trends as well as networking opportunities with industry experts. This continuous educational journey broadens one's knowledge base while keeping skills sharp and relevant.

Certification programmes: Enhancing credibility and expertise

Certification programmes provide a path to increased credibility and specialised knowledge in energy conservation. For example, the certified energy manager (CEM) and leadership in energy and environmental design (LEED) certifications are well-known in the industry. These programmes not only enhance one's understanding, but also demonstrate a dedication to professional excellence. When selecting a certification, it is critical to consider one's career goals and interests. Furthermore, these programmes frequently require ongoing education, ensuring that professionals maintain current knowledge and skills.

Skill development: Essential abilities for energy professionals

Success in the energy sector necessitates a broad skill set. Technical proficiency in energy systems, strong data analysis abilities, and effective project management skills are required. In addition, professionals should improve their communication skills in order to articulate complex concepts clearly and persuasively. Targeted training programmes, participation in industry-specific projects, and collaboration with cross-functional teams can all help you develop these skills. Furthermore, soft skills such as adaptability, creativity, and effective teamwork are critical for developing innovative solutions and promoting energy-efficient practices.

Setting and achieving sustainable career goals

Setting well-defined career goals that are consistent with sustainability principles is critical for long-term success in energy conservation. This entails integrating one's personal values for sustainability into professional goals. Goal-setting frameworks such as SMART (specific, measurable, achievable, relevant, time-bound) can help provide clarity and focus. This dynamic process includes reviewing these goals on a regular basis, adjusting strategies as necessary, and celebrating milestones. This goal-oriented approach not only provides a clear path for professional advancement, but it also guarantees contributions to a sustainable energy future.

Merging personal growth with energy efficiency

The path to becoming a proficient energy-efficient professional combines personal development with professional expertise. It necessitates a dedication to lifelong learning, obtaining relevant certifications, developing a diverse skill set, and setting specific career goals. By embracing these elements, professionals not only advance their careers, but also make a significant contribution to the larger goal of energy efficiency and sustainability.

Being an energy-efficient professional requires a combination of knowledge, skills, and practical steps. Here are six key steps to help you along your journey:

1. **Research international energy regulations**:
   - Familiarise yourself with the energy performance systems and processes required by international standards. These universal standards can be incorporated into the management systems of many organisations and businesses.
   - Notable standards include:
     - **UNE 216.501**: Focuses on energy audits.
     - **UNE-EN 16247**: A European standard.
     - **ISO 50001**: Pertains to energy management systems.
2. **Learn about measuring tools**:
   - When gathering energy data, identify the equipment that uses the most energy in a company.
   - Use specialised instruments to measure flow, temperature, pressure, air speed, electrical parameters, and even gas combustion (for water heaters).
3. **Utilise appropriate software**:
   - Investigate effective energy simulation programmes. Some notable ones (and the good news – they are free!) are:
     - **RETScreen**: An energy analysis tool for financial and environmental improvements.
     - **3EPLUS**: Software for thermal insulation.
     - **EINSTEIN**: Thermal power software.
4. **Conduct thorough reviews**:
   - Evaluate the facilities, technologies, and energy bills.
   - Create a 'energy map' to identify areas with high energy consumption.
   - Determine whether the energy consumption and costs are acceptable.
5. **Analyse and prepare a report**:
   - Determine which aspects of the production process and equipment can be improved.
   - Compare the environmental and economic impacts of various options.
   - Consider measures such as renewable energy sources (e.g. solar panels) or energy recovery techniques.

6. **Remember, no one-size-fits-all approach**:
    - Each company may follow different methods based on their energy resources and context.

13.2 Networking and Collaboration in Energy Efficiency

In the fast-paced world of energy efficiency, networking and collaboration are not only beneficial, but also necessary. Building relationships with peers, experts, and organisations is critical for professionals operating in this sector. These connections pave the way for professional advancement, novel ideas, and collective progress in energy conservation. This section delves into the why and how of effective networking and collaboration, illuminating opportunities for professionals to form meaningful relationships and make significant contributions to the field of energy efficiency.

13.2.1 The power of professional networking

A strong professional network is a valuable asset in the energy industry. It provides access to new opportunities, insights, and resources. Effective networking entails more than simply making contacts; it is about making meaningful connections. Here are a few tips:

- Engage actively on professional platforms: Use tools like LinkedIn to connect with industry leaders and peers.
- Join industry-specific groups. Participate in online energy-efficiency forums and groups to gain insights and discussions.
- Provide value: Networking is a two-way street. Share your knowledge and be willing to help others.

Maximising opportunities at industry conferences

Industry conferences provide excellent opportunities for networking. They are platforms where professionals can learn about the latest trends and network with industry leaders. To make the best of these events:

- Prepare in advance: Investigate the attendees and speakers. Determine who you want to connect with.
- Be active in sessions: Participate in discussions and ask questions during the sessions.
- Follow up post-event: After the conference, contact the connections you made, mentioning specific discussions or sessions.

Collaboration on energy projects

Collaborative projects in the energy sector can result in ground-breaking innovations. They encourage the sharing of knowledge and skills, which would not be possible in isolation. To participate effectively in these projects:

- Seek diverse projects: Look for opportunities that match your skills while also pushing your boundaries.
- Communicate effectively: Clear communication is essential in collaborative projects.
- Be open to learning: Each project provides an opportunity to learn from others and broaden your knowledge.

Building a supportive network for career advancement

A supportive network can be a catalyst for professional growth. It offers mentorship, guidance, and, in some cases, direct opportunities for growth. Consider the following strategies:

- Nurture relationships: Communicate with your network on a regular basis, both offering and asking for help.
- Seek mentors: Look for potential mentors in your field who can offer guidance and advice.
- Attend local meetups and workshops: These are excellent opportunities to meet professionals in a more informal setting and expand your network locally.

Strengthening industry ties for a sustainable future

Networking and collaboration are important factors in the journey of an energy-efficient professional. These practices not only help with personal career development, but also fuel collective efforts to promote sustainable energy practices. As we strengthen our ties within this industry, we help to shape a future in which energy efficiency is not just a goal, but a requirement.

Benefits and strategies for effective networking

1. **Peer-to-peer connections**:
    - Interacting with other professionals and start-ups provides several advantages:
        - Technical exchanges: Work together to solve common problems or look into opportunities to develop complementary products.

    - Information sharing: Learn about each other's needs and share potentially relevant contacts.
    - Business insights: Discuss how to approach business challenges based on recent experiences.
    - Emotional support: Seek solidarity and reassurance from peers who are facing similar challenges.

2. **Investor and potential customer connections**:
    - Networking with investors and potential customers offers:
        - Access to funding: Make connections that may lead to financial assistance.
        - Market insights: Discover your customers' needs and preferences.
        - Partnerships: Look into opportunities for mutual benefit.
3. **Policy connections**:
    - Engage with policymakers and industry regulators:
        - Stay informed: Learn about energy policies, regulations, and incentives.
        - Advocacy: Participate in discussions that shape energy conservation policies.
        - Ensure compliance with legal requirements.
4. **International connections**:
    - Global networking enhances knowledge exchange and innovation:
        - Cross-cultural learning: Learn from various perspectives and practices.
        - Connect with international experts.
        - Collaborative projects: Look into joint initiatives for sustainable energy solutions.

13.3 Energy Efficiency in Personal and Professional Development

In today's world, incorporating energy efficiency into our personal and professional lives is not just an option, but a requirement. With the growing emphasis on sustainable practices, professionals from all fields are expected to adapt and contribute to a more energy-conscious society. This section explores how individuals can pursue a career in energy conservation while aligning their professional development with sustainable practices. By combining personal development and energy efficiency, we can have a positive impact on the environment and set a precedent in our respective industries.

Continual learning in energy efficiency

The landscape of energy conservation is ever-changing, so continuous learning is essential for professionals in this field. Keeping up with the most recent energy efficiency trends, technologies, and strategies is critical. This can be accomplished through a variety of means, including enrolling in online courses offered by reputable universities, attending industry webinars, and participating in workshops. These resources not only broaden knowledge, but also help professionals stay current and competitive. Furthermore, following the blogs, podcasts, and social media channels of energy conservation thought leaders and influencers can provide insights and inspire creative thinking.

Certification programmes and their benefits

Pursuing certification programmes is a practical step towards establishing oneself in the field of energy conservation. Certifications such as certified energy manager (CEM), LEED accredited professional, and BREEAM assessor are highly valued. These programmes not only broaden knowledge, but also boost professional credibility. They demonstrate one's commitment and expertise in energy-efficient practices. Many professionals have reported success stories, with doors opening in terms of career opportunities and advancements following certification. Investing in these certifications often leads to higher responsibility roles, which benefits both personal growth and the larger goal of sustainability.

Skill development for energy-efficient practices

Certain skills are especially valuable in terms of energy efficiency. These include analytical thinking to understand and interpret energy data, innovative problem-solving to address sustainability issues, and technological proficiency to use new tools and solutions. Developing these skills can be accomplished through practical experience, such as working on energy-saving projects or initiatives. Furthermore, seeking mentorship from experienced professionals in the field can provide invaluable insights and speed up skill development. Actively honing these skills not only advances one's career, but also makes a significant contribution to creating more energy-efficient workplaces and communities.

Setting and achieving career goals aligned with sustainable practices

Setting clear, attainable career goals is critical in any profession, especially one driven by the goal of sustainability. Professionals should strive to set goals that not only advance their careers, but also contribute to the larger cause of environmental conservation. This could include pursuing positions that focus on developing green technologies or policies that reduce carbon footprints. Networking within the industry, attending relevant conferences, and participating in sustainability forums can also help you achieve your goals. Regular self-assessment and strategy adaptation are essential for staying on track with both personal values and professional goals in sustainability.

The path forward in energy efficiency

As we work to balance our professional goals with the imperative of energy conservation, it is clear that personal and professional development in this field is not only beneficial, but also necessary. Professionals can help to create a more sustainable future by engaging in continuous learning, pursuing relevant certifications, honing key skills, and setting meaningful career goals. The path to greater energy efficiency is both a personal and a collective endeavour that requires each of us to be dedicated and forward-thinking.

Practical tips for energy-efficient professional development

1. **Staying informed and connected**
    - Regularly read industry publications and join professional associations concerned with energy efficiency and sustainability.
    - Participate in forums and discussion groups to stay in touch with peers and experts in your field.
2. **Leveraging technology for sustainability**
    - Familiarise yourself with industry-leading energy-efficient technologies and software tools.
    - Investigate how digital advancements such as AI and IoT can be used to improve energy efficiency in different projects.

3. **Workplace energy conservation**
    - Encourage and implement energy-saving practices in the workplace, such as optimising lighting and HVAC systems.
    - Encourage colleagues and management to adopt sustainable practices and make energy-efficient decisions.
4. **Community involvement and leadership**
    - Participate in community initiatives aimed at sustainability and energy conservation.
    - Volunteer for or lead projects aimed at reducing carbon footprints in your community or industry.
5. **Continuous personal contribution**
    - Use energy-efficient practices in your personal life to influence your professional approach, and vice versa.
    - Share your knowledge and experiences in energy efficiency with colleagues, mentoring those who are new to the field.

References

[1] *Six key steps to become an energy efficiency expert | IDB Invest www.idbinvest.org*. Available at: https://www.idbinvest.org/en/blog/energy/six-key-steps-become-energy-efficiency-expert. (Accessed: 08 May 2024)

[2] Reyes, M. delos (2023) *Energy Consultant: A Complete Guide to the Career, Earnings, and Business - Sustainability Education Academy*. Available at: https://sustainabilityeducationacademy.com/energy-consultant-a-complete-guide-to-the-career-earnings-and-business/. (Accessed: 08 May 2024)

[3] Energy Technology Innovation Partnerships – Analysis - IEA (2019) *Energy Technology Innovation Partnerships – Analysis - IEA, IEA*. Available at: https://www.iea.org/reports/energy-technology-innovation-partnerships. (Accessed: 08 May 2024)

[4] *Adapting the Energy Sector to Climate Change, IAEA Publication Available* (2019) *www.iaea.org*. Available at: https://www.iaea.org/newscenter/news/adapting-the-energy-sector-to-climate-change-iaea-publication-available (Accessed: 8 May 2024). (Accessed: 08 May 2024)

14

Conclusion

Final Thoughts and Encouragement for Action

As we conclude this book, we recognise that the path to sustainable energy and conservation is both collective and personal. Each professional in their respective field possesses the ability to influence and implement change. This final chapter serves as both a reflection on what has been accomplished and a rallying cry for continued action and dedication.

14.1 The Role of Professionals in Shaping a Sustainable Energy Future

Engineers, policymakers, energy consultants, and others form the foundation of initiatives aimed at promoting energy conservation and managing the transition to more sustainable practices. This final section considers how the collective efforts of these professionals are not only critical, but also transformative in addressing global energy issues.

Engineers: Pioneers of technological innovation

Engineers are at the forefront of developing technologies critical to renewable energy systems and energy-efficient solutions. Engineers create solutions that make sustainable energy viable and cost-effective, from advanced solar panels and wind turbines to improved battery storage systems.

- Engineers play a critical role in the design and deployment of renewable energy systems such as solar, wind, hydro, and geothermal.
- Smart grid technology: Electrical and system engineers create smart grid technologies to improve energy distribution and reduce waste.
- Energy-efficient building designs: Civil and architectural engineers incorporate energy-efficient technologies into new building designs, resulting in significant energy savings.

Policymakers: Setting the framework for sustainability

Policymakers have the unique ability to influence public and private sector behaviour through laws, regulations, and incentives. They establish legal frameworks that encourage or, in some cases, mandate the use of renewable energy and energy-saving measures.

- **Regulatory policies**: Implementing policies that require increased use of renewable energy sources and reduced carbon emissions.
- **Incentives for renewable energy**: Establishing tax incentives, grants, and subsidies to promote renewable energy investments.
- **International agreements**: Facilitating and enforcing international agreements that aim to reduce global carbon emissions and promote energy efficiency globally.

Energy consultants: Advisors for best practices

Energy consultants play an important role in bridging the gap between current practices and more efficient, sustainable approaches. They assess energy usage patterns within organisations and make recommendations to improve efficiency and reduce environmental impact.

- Energy audits and assessments involve conducting thorough energy audits and developing strategies to address inefficiencies.
- Sustainable energy planning: Assisting organisations in incorporating sustainable energy solutions into their operations and long-term plans.
- Training and education: Providing companies with training on how to implement and manage new energy systems and practices.

14.2 Collective Impact and Innovative Approaches

The combination of these professionals' efforts has a synergistic effect, propelling global progress towards energy sustainability. Collaborative projects and interdisciplinary approaches greatly increase the potential for innovative solutions to energy challenges.

- Collaborative innovation: Engineers, policymakers, and consultants work together on projects that combine multiple perspectives to provide holistic solutions.
- Technology and policy integration: Ensuring that new technologies are supported by strong policies that encourage their adoption and integration into the energy market.

- Public awareness and engagement: Professionals from all sectors work with the public to educate and promote energy conservation and sustainability.

The path to a sustainable energy future is complex and difficult, but also extremely rewarding. It necessitates the dedicated participation of professionals from various sectors, each contributing their expertise to solve a piece of the puzzle. By making informed decisions and implementing innovative approaches, these professionals not only help to solve immediate energy challenges, but also help to shape the path to a sustainable and resilient energy landscape. As we move forward, the collective efforts of these individuals will continue to drive the transition to a sustainable energy future, demonstrating the profound impact that skilled and dedicated professionals can have on global energy solutions.

The power of individual action

The global shift towards sustainable energy systems and practices is dependent not only on policies and technologies, but also on the individual actions of professionals such as you. Every small step towards energy efficiency leads to a larger, more significant change. Whether you are an engineer, policymaker, consultant, or other professional, your decisions and innovations will have a significant impact on our energy future.

Encouraging proactive steps

1. Continue learning: The field of energy conservation is constantly evolving. Commit to staying current on the latest technologies, trends, and policies. Attend workshops, take courses, and read extensively to keep your knowledge up to date.
2. Foster innovation: Encourage and participate in innovative projects within your company. Experiment with new ideas and technologies that have the potential to improve energy efficiency or reduce reliance on non-renewable energy sources.
3. Lead by example. Set a high standard for energy efficiency at your workplace. Implement sustainable practices in your operations and encourage your co-workers to follow suit. Leadership in this area has the potential to have a far-reaching impact.
4. Advocate for change: Use your expertise to promote sustainable energy policies and practices in your community and elsewhere. Influence can

be gained through writing, speaking, or serving on governmental or advisory committees.

5. Collaborate across disciplines: Work with professionals from other fields to develop interdisciplinary approaches to energy challenges. Collaborations can result in breakthroughs that a single discipline may not be able to achieve on its own.

Encouragement for continued engagement

The journey to energy sustainability is ongoing and requires the continuous effort of dedicated individuals. Remember that your work has an impact not only on the present but also on the world that future generations will inherit. Let this serve as motivation to strengthen your commitment to promoting energy efficiency and sustainability.

You are a part of a global movement – a community of change-makers committed to keeping our planet liveable and vibrant for generations to come. Your expertise, enthusiasm, and involvement are critical to this cause. Let us forge ahead with a renewed commitment to innovation, leadership, and proactive change.

Allow your efforts in energy conservation and management to be motivated by a vision of a sustainable world. Let each decision and innovation be a step towards increased energy efficiency and a healthier planet. You have the resources, knowledge, and ability to make a difference. Step forward with confidence and purpose, and become the change you want to see in the world.

Index

About the Author

Eng. Benard Makaa, is a professional engineer and a class A1 electrician. He holds a B.Sc. and M.Sc. in Electrical Engineering. He works as a researcher, consultant and lecturer in Electrical Engineering. He also has extensive experience as a consultant in electrical building services, and has authored a book on the same area: "Electrical Services for Buildings: A Consultant's Guide". More details about his work and professional affiliation can be obtained at www.benardmakaa.com.

His other published works:

1. *Electrical Services for Buildings: A Consultant's Guide*
2. *Smart Way to Improve your Mind and Grades: A Revolutionary Plan to Get into University by Standing Out (Without Burning Out)*